THE WORLD AS WE KNEW IT

THE WORLD AS WE KNEW IT

DISPATCHES FROM A CHANGING CLIMATE

EDITED BY

AMY BRADY AND TAJJA ISEN

CATAPULT NEW YORK

ISBN: 978-1-64622-030-4

Library of Congress Control Number: 2021947150

Cover design by Nicole Caputo
Book design by Wah-Ming Chang

Catapult
New York, NY
books.catapult.co

Printed in the United States of America
1 3 5 7 9 10 8 6 4 2

To those around the world
who advocate a just and sustainable future

CONTENTS

INTRODUCTION

Summer nights in Kansas used to be so alive with the sounds of insects and animals that, if you wanted to tell a story, you had to shout to be heard. The trees vibrated with the droning of locusts; the weeds shook with crickets. Frogs bellowed up and down the creeks. The Brady family spent most summer evenings in the backyard, where we sat on torn folding chairs beneath a flickering porchlight, hollering at one another to be heard over the night's wild buzzing.

The kids, myself included, would chase fireflies. In those days, the darkness out in rural Kansas was still complete, an inky black pierced only by the porchlight and the blinking bugs and the headlights of the occasional car that zoomed down the distant I-70. In all that darkness, the fireflies were a cloud of light. When they swerved so did I, an open glass jar held tightly in one hand, a metal lid with three tiny holes in the other. Once, I tripped on a twig and went crashing down into the grass. Knees burning, I rolled over and stared up at the night sky, the fireflies and stars so thoroughly blended that I couldn't tell when a stream of light was a bug or a shooting star. It never occurred to me that nights like those wouldn't always be possible.

These days, summers in Kansas have changed. Thermometer readings now regularly reach the upper nineties. Corn and milo crops are suffering, and some people without air-conditioning are routinely hospitalized with heat stroke.

Now a New Yorker, I recently visited my family back in Kansas. The old house has been sold, but my father still lives in Topeka. The silence of the animals and insects on this trip was unsettling. Owing to the development of more grocery stores and restaurants, the darkness was gone as well. Here was a landscape so familiar I could draw a map from memory, and yet it felt strange to me.

That peculiar feeling has become recognizable to many people around the world. In 2022, we are witnesses to one of the most transformative moments in human history: a time when climate change is altering life on Earth at an unprecedented rate, but also a time when the majority of us can still remember when things were more stable. We are among the first—and perhaps one of the last—human populations to have memories of what life was like *before*. To us, the "new normal" is not yet how it's always been. Our lives jostle against incongruous memories of familiar places. We are forced to confront, in strange and sometimes painful ways, how much those places have changed.

When we think of environmental crises our minds might go first to extreme weather events, like Superstorm Sandy, whose size and scale were amplified by climate change. At approximately 8:00 p.m. on October 29, 2012, Sandy struck Atlantic City, New Jersey. That night, a full moon hung in the turbulent sky, pulling the ocean tides a full 20 percent higher than normal and increasing Sandy's storm surge. Seawater rose along the Eastern Seaboard toward New York City and then poured into Manhattan, flooding subways and sidewalks. More than six hundred thousand people lost power throughout the five boroughs; many would be without electricity for more than a week. The years since Sandy have seen an escalating series of even larger events. In 2020, there were so many tropical storms that the World Meteorological Organization nearly ran out of names for them.

But the connections between humans and the natural world go beyond extreme weather events. As the Earth warms, other devastating phenomena continue to thrash the planet: invasive species migrate to cooler climates, choking off local wildlife and creating potentially threatening moments of contact between animals and humans. Wildfire "seasons" are now year-round. Low-lying nations threatened by sea-level rise, like the Marshall Islands, are being forced to consider a terrifying, almost inconceivable choice: relocate the entire population or elevate the land. For the Marshallese, the latter would involve raising 1,200 islands scattered across 750,000 square miles of ocean. Early in 2020—and partway through compiling this anthology—the COVID-19 pandemic encircled the world, altering our lives in ways that are by now familiar. The catastrophic novel coronavirus was borne out of humanity's complex and unsustainable relationship with wildlife. The way things are headed, this pandemic likely won't be the last.

To use a metaphor that has grown uncanny, these visible effects of global warming are just the tip of the iceberg. In the public conversation about climate change, macro change tends to take center stage, and for good reason: it impacts the lives of millions and serves as an increasingly urgent reminder of the need for decisive action. But less told among the literature of climate change are the stories of individuals—how they're coping (or not) with the changes occurring in their own lives. That's the scale that *The World as We Knew It* seeks to highlight—not by turning away from global events, but by emphasizing the links between the individual, the collective, and the environmental. Sometimes the connections between the personal and the planetary can be hard to see, but once we start looking, we notice that they're everywhere.

"At a time when our planet is experiencing terrifying and unprecedented levels of change, what corresponding transformations have you witnessed in your own lives, yards, neighborhoods, jobs, relationships, or mental health?" That's the question that Tajja Isen and I asked our contributors to this anthology. We wanted to hear their personal stories, allow them to serve as witnesses of this increasingly complex moment in history. We encouraged them to take the theme in any direction they desired, and indeed, they did.

Some of the pieces were written and edited before the appearance of COVID-19 and reflect a world that hadn't yet been altered by it. For other pieces, the pandemic erupted partway through the editorial process, requiring editors and contributors alike to consider new approaches to storytelling that could account for yet another kind of global upheaval. Others still were assigned after our world had irrevocably changed, and were drafted and revised through civil unrest, a harrowing election, insurrection, police violence—an absurd number of converging crises that demanded our attention and commitment. Out of those various disruptions, though, *The World as We Knew It* has become a kind of living document—a record of things as they were, a testament to living and writing through tragedy, and an exercise in envisioning the life that might await us on the other side. The events of the past few years have forced us to consider an entirely new set of connections between the individual and the global. We're grateful to our contributors for persevering with this project alongside us.

The works collected in *The World as We Knew It* reflect these various states and times as they explore the relationships between humanity and our environments. Emily Raboteau's essay, "How Do You Live with Displacement?," is a polyvocal record of the first three months of 2020, charting the dual threats of climate change and the novel coronavirus. Raboteau chronicles the voices of her community when New York City was the epicenter of the U.S. outbreak, shedding light on the virus's disproportionate impacts on Black and brown people—a form of inequality replicated by the effects of climate change. Porochista Khakpour's "Season of Sickness" traces the connections between an ailing climate and human diseases, especially the ways that the former can aggravate the latter. As Khakpour struggles with her own Lyme disease, trying to find a home that doesn't worsen her condition, the pandemic casts its first shadow in the United States. Meera Subramanian's essay, "Leap," set primarily on Cape Cod, focuses on a similar intersection of climate change and sickness through the figure of the tick. Subramanian poignantly illustrates how a growing awareness of the natural world's dangers can alter our once-idyllic relationship with it.

Other writers figure their relationship to the planet through lenses like nuclear testing, anti-Black racism, and international travel. Pitchaya Sudbanthad's "A Brief History of Breathing" takes place in Bangkok, Saudi Arabia, and New York City in the immediate aftermath of 9/11—a haunting triptych of air pollution. In "Walking on Water," Rachel Riederer brings to life a debate between two indigenous leaders in Uganda over how best to move ancestral spirits living at the construction site of a new dam. The essay drives home how colonial development alters more than just ecosystems—it destroys the homes and histories of indigenous people. Lidia Yuknavitch's "Unearthing" offers a chilling account of the Hanford Site in Washington State, a longtime site of violence done to the land—and its generations of inhabitants—by nuclear production, weaving its histories of destructive experimentation with vignettes of childhood life. In "Iowa Bestiary," Melissa Febos charts her changing relationship to her new home through interactions with its wild inhabitants. Taking us to the Arizona desert, Lydia Millet elegizes the plants and animals of Saguaro National Park, describing the bittersweet sense of loss that accompanies putting down roots in the midst of a declining ecosystem. Mary Annaïse Heglar points out that Hurricane Katrina eerily coincided with the fiftieth anniversary of Emmett Till's murder. She goes on to describe how Katrina shaped her understanding of the connections between environmentalism and racism in the United States.

In other essays, contributors explore how their personal histories continue to shape their contemporary understanding of climate change. Lacy M. Johnson's incantatory "Come Hell" finds surprising links between childhood memories of church, floods, and coal mining. Remembering her family home in Dominica, Gabrielle Bellot ponders the arrival of an invasive fish species before considering the tiny animal's planet-sized connection to Hurricane Maria. The novelist Omar El Akkad writes about how the Persian Gulf—his childhood home—is no longer recognizable to him, and how, as a writer, the obliteration of his home means he's also losing the wellspring of his art. Tracy O'Neill remembers fondly a childhood spent in New Hampshire, only to wonder now whether bringing a child of her own into the world makes moral sense. Nickolas Butler likewise considers the impact of climate change on young people. Thinking of lessons he learned from his disaster-prepper father, he wonders what to impart to his own children to prepare them for a climate-changed future.

Wonder is another recurring theme in this collection. In "Signs and Wonders," the Australian writer Delia Falconer compares life in an age of climate change to that of ancient Rome, when priests would look to wild animals and the weather for clues about what the future might bring. Alexandra Kleeman writes about falling in love with a most surprising sight—an apple tree on a slip of green near her newly purchased home on Staten Island—only to learn of its planned removal to make way for a new shoreline development. In "Cougar," Terese Svoboda is motivated by the alarming sight of a wildcat on a highway to learn more about how climate change has reshaped the bounds of the animal's territory. What she learns forces her to consider how an altered world might also shape the future of her family. In "Moments of Being," Kim Stanley Robinson writes about three camping trips he took with friends in the Sierra Nevada, during which he experienced firsthand the awesome power of extreme weather—weather rarely seen in that range before our climate went haywire. Pondering the wondrous sight of the Thwaites Glacier in Antarctica, Elizabeth Rush reminds us that women weren't welcome during the continent's first two hundred years of exploration. Only now, at a time when the continent is hemorrhaging ice, are women allowed to visit what's left.

Taken together, the essays in *The World as We Knew It* create a timely, haunting mosaic of life in the age of climate change. They also emphasize that the most astonishing transformations happening on our planet are the result of our own actions. Scientists have known about human-driven climate change since the late nineteenth century, and politicians have been aware of it since at least the 1970s. But despite these warnings, neither governments nor large corporations have taken aggressive action. Whether we act more responsibly in the future remains to be seen, but for now, the changes continue. By turning their focus inward, our contributors reveal the psychological impacts of climate change, an emphasis that Tajja and I hope will encourage a humble and humane dialogue; one that inspires reflection on the changes one can make at the level of a life. Given the scale of our present situation, we understand that individual experiences like the ones described here aren't the most intuitive ways to think about climate change. But we believe they are among the most powerful.

Amy Brady and Tajja Isen
Co-editors
2022

THE WORLD AS WE KNEW IT

FROM THIS VALLEY, THEY SAY, YOU ARE LEAVING

LYDIA MILLET

I'm devoted to where I live—some might say too devoted, since I've left people and jobs elsewhere to stay out here. Once, after a local utility sent me a letter announcing it planned to blade a swath across my patch of desert to lay some cable, I dissolved into a fit of sobbing. That was fifteen years ago, and so far the bulldozers haven't come: from what I can tell, they chose to blaze a shorter trail through the property next door instead.

Which happens to belong to my mother. She replanted a few bushes and cacti along the path they plowed, but you can still see, in the bare ground around those sparse plantings, the footprint of the machines.

It's not that I experience the "pride of ownership" that real estate agents like to tout in ad copy. The phrase doesn't resonate with me, no matter how hard I reach for a feeling that might approximate it. I'd like to think ownership meant something beyond resale value—say, permission to protect a place from harm—but mostly, as far as I can tell, unless you're willing to lie down in their path, the bulldozers will come whether you own a place or rent it. It's far truer to say the land owns me: it has all the power in our relationship.

Somehow it became my only right home as soon as I set foot here. I can't say exactly why, because my long, straight street, as you drive down it toward the turnoff for my house, looks almost ugly. I see prettier streets all the time. But as soon as I head up my winding dirt driveway, I find myself in a lush tumult of trees and cactus and shrubs—lush to the eye, though piercing to the touch. With the angle of the sloping land against the sky, and without the light pollution of Tucson on the other side of the mountains, I can sometimes stand out in my yard in the darkness, look up, and see not only the constellations but the wide, hazy path of the Milky Way.

The expanse of stars, the multiple mountain ranges visible from my balcony, the cactus and other otherworldly, curious forms of plant life all come together with a singular beauty. Wild animals wander through my yard: mule deer, bobcats, and collared peccaries that look like boars, which we call javelinas. Foxes, coyotes, roadrunners, quail. Others alight on trees: owls, hawks, flickers, songbirds. Still others live down in the coolness of the earth, like the round-tailed ground squirrels that remind me of miniature prairie dogs, Harris's antelope squirrels that resemble chipmunks, many lizards, tortoises, and toads.

My home lies adjacent to Saguaro National Park, whose low but dramatically pointed volcanic mountains are spiked with tall saguaro cacti, the iconic plants of the American Southwest you used to see in cartoons about the desert and cowboy movies. They're often thirty or forty feet tall, with arms that make them look creaturely. Spread out and upraised, as though they're constantly hailing you in greeting.

There are other wondrous cacti as well, some in the shapes of paddles, others like barrels, still others like clumps of breasts growing out of the ground or trees with thousands of spiny segments that get lit up gold by the setting sun. Woody plants called ocotillos with red flames for flowers, thorny trees with green bark.

Still, it's the saguaros that most define the Sonoran Desert. These giants can weigh more than five thousand pounds and, in good conditions, outlive people by dozens of years and sometimes centuries—one saguaro on the east side of the city, called Granddaddy, was three hundred by the time it died in 1992. They're a keystone species that many birds depend on for nesting, from woodpeckers to tiny elf and screech owls. Bats, bees, and birds pollinate them. Javelinas, coyotes, and tortoises eat their large, fuchsia fruits. As do the Tohono O'odham, who've lived here for thousands of years. They make jelly from the fruit, along with wine that's consumed during rain ceremonies.

But the so-called megadrought that has descended on the West since 2000 ranks among the greatest droughts of the past millennium. The animals that live here are hardy, but their water-saving adaptations depend on a fragile balance—the desert's green things grow and reproduce with painful slowness. My mother, who moved here a few years after I did, has rain gauges stuck into the ground across her land, tiny beakers with wire frames. She'll shake her head after a particularly dry summer monsoon season, telling me sadly how many inches we got, as though each fraction of an inch is gold.

And it is—better than gold.

Eventually, they say, the changing climate may plunge our region into conditions reminiscent of the Dust Bowl.

Before I moved here I didn't tend to stay in the same place for too long—three years maximum, as an adult, before I found this desert when I was thirty. Admittedly, I've traveled a lot: my contentment and career both depend on the connections offered by cars and airplanes, which break up my isolation and let me see friends and visit family. So I'm a conservationist who worries about carbon all the time but can't live in her private haven without leaving a heavy footprint.

My life isn't what you'd call sustainable. It's a rural life, of a sort, since the only businesses out here are gas stations, a couple of dollar stores, and a dive bar called the Wagon Wheel Saloon. But it's not a farming life. Unless you give yourself over to cultivation pretty much full-time, and don't mind sucking plenty of water out of the river system and overdrawn local aquifers, the desert's not a great spot for food growing. I don't grow anything I eat save rosemary from a single bush in my garden, one that's apparently impossible to kill.

Not everyone has found this desert so difficult to cultivate. In fact, southern Arizona has been continuously inhabited for longer than most places in the country—more than ten thousand years. People are thought to have lived here in Avra Valley in the Middle Archaic period, between 5000 and 1500 B.C., and the

Hohokam were here by 200 B.C. The Pascua Yaqui were in the area by around 550 B.C. and thrived for many centuries before white people like me arrived—with no access, that we know of, to either grocery stores or restaurants.

These days my town is a satellite of a city that is itself a satellite—of the Colorado River, which now provides, through a multistate water-sharing compact, much of its water. My routine and job depend on internet technology without which, on top of my car and its gas, and that water that's piped in from far away, I couldn't feed myself or my children.

Our living here looks more like suburban luxury than rural self-sufficiency—a gift of fossil fuels.

When I bought my house, it was a decrepit structure thrown together without permits, probably first in the 1950s, though the earliest county records are from twenty years later. Made mostly of army-surplus plywood, it featured studs that were often placed at an angle no self-respecting builder would tolerate—not perpendicular to the sheetrock but parallel to it. The floors were peeling burgundy linoleum, and the walls lined with that 1970s fake-wood paneling that doesn't fool anyone, also made of vinyl. The house was so porous, when I first moved in, that black widows and bark scorpions—*Centruroides sculpturatus,* the only lethal scorpions in North America—had colonized the interior. One of these crawled into my bed while I was sleeping and stung my hand twice, where it lay under my pillow.

Termites and pack rats had taken up residence inside—the latter, at one point, accumulating hundreds of pounds of cholla cactus, ripped-up insulation, random-seeming trash, and of course their feces in the massive nests they built in the walls and crawl space. Western diamondback rattlesnakes lived in holes all along the perimeter, where the dirt met the concrete slab. Still do.

I replaced the roof and raised the ceilings, replaced the linoleum with the Saltillo tile that's typical of Mexican Sonoran homes. I built an addition onto the house, installed a solar array and solar water heater, gutted the main living space, and cut arched doorways between rooms. Planted native vegetation where the previous owners had left a barren field of dirt surrounding the dwelling and installed a drip system that doesn't waste too much water for irrigation. I had a pond built, where fish live and lily pads grow and wildlife come to drink. The fish and the lilies are non-native, but since the nearest other water is likely a dog bowl or stock tank a half-mile away, there's not much chance that they'll spread.

I've never been drawn to the cowboy myth, whose romance of independence belies the fact that livestock operators' off-road-vehicle-driving lifestyle is neither romantic nor independent. Beef growers have done far-reaching harm to the West, their cattle trampling and polluting rare, precious rivers and streams and driving a range of native plants and animals extinct. And they're heavily subsidized by taxpayers.

But I do love the solitude that myth evokes—along with many other desert myths, from the ancient holy books on—as a form of meditative communion between the small self and the limitless universe, the finite and the infinite. The music and noise of people and the silent majesty of open country.

And I love old, sad cowboy songs. One of them has been running through my head lately, possibly because my daughter had to memorize a verse of it, translated into Mandarin, to perform with a group on a school trip she made to China last year. The English lyrics, sung by Marty Robbins in the rendition I know, go like this:

> From this valley, they say, you are leaving
> We will miss your bright eyes and sweet smile
> For you take with you all of the sunshine
> That has brightened our pathway a while.

In my part of Arizona, we're already seeing changes in seasonal weather patterns that seem to confuse the animals and the plants, bringing them out of hibernation at odd times or into bloom when the right pollinators aren't around. The summer monsoon, which so many wild things depend on to survive, delivers about two-thirds of southern Arizona's annual rainfall between July and September. But its storms are growing more frequent and intense, and heavy rainfalls result in less water soaking into the ground—more of it ends up as runoff. At the same time, average daily rainfall appears to be decreasing. So less rain is falling, while more powerful storms are reducing the availability of water to wildlife and people alike.

And the saguaros, like other native species, are suffering. Long-term surveys, conducted by volunteers who've been faithfully counting cacti for half a century, show that on the east side of town there aren't enough young ones rising from the ground.

In some places, there's hardly any new growth. When the old ones die, they're not being replaced.

Even if it turns out that the towering cacti are not yet in decline, as other observers have argued—here on the west side of town they seem to be doing better—climate change means they'll be increasingly at risk from man-made conditions made more severe by the rising heat and drought. An African plant called buffelgrass, introduced for cattle to eat, is chief among the villains, since it promotes wildfire in an ecosystem not evolved to regularly burn.

In the future, higher daytime and nighttime temperatures are projected to reduce saguaros' water-use efficiency, making them more likely to die during drought periods. They can tolerate occasional brief freezes, though they don't like prolonged subzero spells: in 2011 a catastrophic freeze killed off many older saguaros. The species rarely grows at elevations above four thousand feet.

On the other hand, as freezing temperatures become rarer, saguaros may be at a disadvantage from that side, too—outcompeted by exotic vegetation that fares better when it doesn't freeze.

I've seen several giants die on my land, among them one with dozens of arms that was likely a couple of centuries old. This cactus used to loom over the end of my driveway, near my house, but fell suddenly during the long, hard freeze of 2011. More often they're killed in another way: a black ooze called bacterial necrosis spreads from one of the bird holes in the cactus, emitting what some call a foul odor. I don't find it foul, but it's certainly strong. As the disease spreads, the cactus rots, turns black across large portions of its trunk, and finally collapses—sometimes in segments, dropping one or two arms at a time, and sometimes all at once.

There are other looming threats to the saguaros and the wildness that remain in this valley. Some more imminent, even, than runaway climate change. Among them is a planned freeway bypass that would cut straight through the valley, between the national park on one side and a national monument on the other. A freeway would block animal migrations, kill animals directly, and bring noise pollution and dirtier air and water.

No one in the area wants it—it would be more expensive and far more destructive than simply widening the part of Interstate 10 that already runs through the western part of the city, on the urban side of the Tucson Mountains—but state and federal transportation agencies are pursuing their "preferred alternative" (the ruin of our valley) with dogged perseverance. Despite resounding public opposition.

I've lived in this desert long enough to have buried several animals in my yard. A dog, two cats, and one coyote. A small group of stones bears the pets' names. I used to imagine that I, too, would be able to grow old and die in the place I love most. I don't *wish* to die, needless to say, but it once comforted me to think it might happen here. And still does, when I'm feeling hopeful about the future.

I like to believe I won't have to dwindle into frailty hobbling along a treeless sidewalk in some gray winter in New York, beside garbage cans and parked cars. That instead I may be allowed to grow old in the desert. Maybe even manage to be outside when the moment comes. Slip away under the stars, as the warm wind moves the branches of trees.

Of course, no luxury is greater than being able to choose the manner and place of your own death. And more and more it seems to me I may not be granted that luxury. Because if the desert begins to die too visibly—if I see its native life turn brown, as the drier, hotter seasons pass, and vanish around me—I know I won't be capable of staying. I know I'll have to retreat. It would break my heart too slowly to watch the decline of what I hold so dear: a death by a thousand cuts. I'd need to flee, abandoning my home before the worst ravages hit.

A part of me waits to see whether I'll have to take the coward's way out. I'd feel a terrible guilt in the act of abandonment. But guilt is more bearable than the pain of loss. So I wait to see whether too many of the giants begin to crumble and collapse onto the sand, their curving arms no longer raised to greet us.

Or whether, through the saving grace of social pressure and political will, a miracle will occur—the one we need to summon to have a chance of staving off the many catastrophic effects of a swiftly changing climate. In my home and far beyond. That will flatten the curve of extinction, slow the destruction of the natural world.

And if, through that grace, the desert will show signs of living on after I'm gone.

STARSHIFT

GABRIELLE BELLOT

In 2011, shortly after the start of the new year, an unusual creature appeared in the waters around my island. Unlike the legendarily vast boars my father claimed he had hunted with his father in the previous century in the primeval reaches of Dominica's mountains, or the witch-women we called *soucouyant* who were said to abandon their skins at night and fly across the skies as balls of firefly flame in search of children's blood to drink, *this* little creature was undeniably real. A specimen had been caught by Dominica's Fisheries Division.

The fish's slim bronze-and-white-striped body was barbed on the top and sides with venomous, striated spines; its eyes were wide and gloomy—the kind of unnerving yet elegant entity one might expect to find in the aquarium of a conniving Bond villain. It was a red lionfish, the first of its kind recorded in Dominica, as striking and strange as a waterfall in a desert.

At first, it was just a curiosity, if not a conspiracy, for those Bible-quoting Dominicans who imagined that all new things were the result of Satanic forces gripping the planet. I was intrigued by its peculiar, perilous beauty. While scuba-diving, I loved floating just beyond the sea's extraordinary creatures, even the ones I knew might hurt me if I got too close: the blue morays that poked their heads out of rocks with mouths open, the purple and black sea urchins with their spines jiggling in the current, the brown stingrays that glided like great magic carpets over the sand. The lionfish would simply be yet another marvel of nature to admire from a distance.

The lionfishes began showing up more and more, filling the excited stories of grinning snorkelers and the nets of flummoxed fishermen, who scratched their heads as they pulled their blue and red dinghies ashore to the beaches. Chefs began speculating that they had a new delicacy on their hands. Dominica's food was delicious, but we were not well known for culinary innovation; a new fish on menus seemed exciting.

Soon, people stopped smiling. The lionfish was everywhere. Turn on the radio or TV news, and people were talking about the spiny creature. It had quietly begun to colonize our reefs, disrupting the ecosystems through the voracity of its appetite. As it spread, other creatures left, or died off, for it had few predators. As the lionfishes' numbers swelled, the coral reefs we once expected to be there forever for us to explore, our bodies surrounded by shimmering clouds of fish, were slowly emptying of other marine life, evanescing away. An atmosphere, it seemed, had shifted in Dominica, like a party whose soundtrack had gradually moved from soca to the blues.

It was strange, this invasion happening in a country that had already been colonized multiple times. But unlike the French and the British, the lionfish represented a slower, more serpentine takeover. It was an altering of our world, not through brute force, but by the erosion of its ecosystems.

And it wasn't just in Dominica. Across the Caribbean, reports abounded of lionfish invading our reefs in troubling numbers. In North America, scientists began speculating that as climate change caused colder waters to gradually become warmer, invasive tropical species like the lionfish might expand their reach further still.

We became accustomed to the new situation. People were advised to catch and kill them on sight; they were now less a novelty than a staple on certain restaurants' menus. Diving companies began focusing their scuba locations on where they could find the most lionfish to hunt, even if it meant taking customers to the same sites over and over. Over time, the invaders' numbers dwindled, even as they remained a threat.

But why they had come in the first place was a mystery few people, myself included, wondered too deeply about at the time. It was just a thing that had happened, a ripple on the often-still lake of life on a small island. And when you feel too accustomed to stillness, you yearn for those ripples, big or small.

I didn't realize until years later that it was the kind of strange thing that would accumulate in a long line of other curious events, all related, ultimately, to our planet's fluctuating climate norms. The kind of *strange* that, over time, we would acclimatize to and tell ourselves was actually the way things had *always* been, even if such rationalizations were indisputably untrue. It was as if a star's brightness in our sky had slowly shifted, night after night, until we came to believe, astronomy notwithstanding, that it had always been that intense, and we were just wrong before.

But then again, humans are stunningly resilient. Even after

trauma, we keep going, though the climates of our world and of our selves have changed. Perhaps we are better at noticing changes in ourselves—the sudden snowfalls in the hallways of our hearts, the ghostly gusts in our corridors from old pains reasserting themselves, chilling us, stiffening our limbs, making us panic in public—than we are at seeing those in the climate of our planet.

What is the era of the human, the Anthropocene, if not an era where our telescopes are sharper than ever, but we choose not to see?

Like the townspeople in Camus's *The Plague*, many of us in Dominica turned our heads the other way, though it was clear that some change was afoot. As in much Anglophone Caribbean literature, we had an ingrained, if not always voiced, pessimism about change in our island, a *so-it-goes* sense that things would always more or less remain the same, rather than changing for the better. We believed that, like the ever-reappearing potholes in our roads, this issue with a new fish would just work itself out somehow, and soon we would return to our eternal gripes about the corruption of our prime minister and the seemingly unending fall from grace of our region's once-great cricket team.

Regardless of one's position, the lionfish had become a symbol. Our world was transforming—and it was getting harder and harder to ignore. Something subtle about the fabric of our day-to-day reality, our expectations, had shifted, and, through a hole in that contorted fabric, something small but bright had swum into a place it had not been before.

Perhaps, in retrospect, it was the presaging of the great storm that would come, years later, when too many of us still had our eyes closed.

•

In 1969, the Martinican writer Aimé Césaire released a striking adaptation of—or, really, response to—*The Tempest*, perhaps the most racially charged of Shakespeare's plays besides *Othello*. Césaire's version, *Une Tempête*, follows the basic premise of Shakespeare's play, leaving much of the world the same—the setting and characters are "as in Shakespeare"—yet this world is also irrevocably different, partly because Césaire specifies in an addendum that the spirit Ariel is "a mulatto slave" and Caliban is "a black slave." By implication, Prospero, the Milanese duke-magician shipwrecked off an island that appears to be in the Caribbean, is a white slavemaster.

These choices reflected Césaire's desires for *Une Tempête* to exist as a blunt example of Négritude, a far-reaching literary-political movement adopted by Black Francophone artists and intellectuals, which sought to critique and condemn colonialism and to lionize Blackness rather than whiteness.

In the Bard's drama, Prospero, shipwrecked on an island with a motley crew, decides to enslave a nonwhite resident of the island, Caliban, whom he treats as a piece of despised property. In Prospero's eyes, Caliban, the child of the infamous witch Sycorax, is simply a monstrosity, a "freckled whelp hag-born—not honour'd with / A human shape." The other white characters react similarly to Caliban, dehumanizing him at every turn. Trinculo, the jester of the king of Naples, is stunned when he comes across Caliban's sleeping form, asking himself in alarm whether Caliban is "a man or a fish," and frequently describes him, as does Stephano, the king's butler, as a "monster." Caliban is cursed and spat upon; he and Ariel become the symbols of the colonized being demeaned and demonized by their European colonizers.

Caliban sometimes speaks back to Prospero—"you taught me language," he says to Prospero, "and my profit on't / Is, I know how to curse"—and Césaire amplifies this, giving Caliban many mocking, indignant lines. "With that big hooked nose, you look just like some vulture," he insults Prospero soon after his first entrance. Caliban's speech, both angry and intentionally humorous, makes it clearer that he may be of African descent; for instance, when he appears in a scene midway through the play, he announces his presence by saying "Uhuru!"—a Swahili term for "freedom"—which was well known amongst Pan-Africanists as an exclamation evincing one's right to liberty as a Black person. Through the power of a voice allowed to burst forth with mad, mocking force, the atmosphere of Shakespeare's imagined island changes utterly in this adaptation.

A voice, Césaire knew, can conjure a storm, a reweaving of a land. In his surreal masterwork, the long poem *Notebook of a Return to the Native Land*, Césaire describes the process of learning to invoke nature through language. Speaking the right way, he suggests, is a kind of magic, an incantation that brings *word* and *world* together:

> I would rediscover the secret of great communications
> and great combustions.
> I would say storm. I would say river. I would say
> tornado.
> I would say leaf. I would say tree.
> I would be drenched by all rains, moistened by all dews.

I would roll like frenetic blood on the slow current of the
eye of words
turned into mad horses into fresh children into clots into
curfew
into vestiges of temples into precious stones remote enough
to discourage miners. Whoever would not understand me
would not understand any better the roaring of a tiger . . .
I would have words vast enough to contain you and
you earth taut earth drunk

The right language, he says, alchemizes sound into the thing itself. I can make storms with words, words from storms.

I pause as I write this and wonder, frowning, about tempests. Tempests are natural, a destructive part of an ecosystem, like wildfires in California. Yet our tempests in the Caribbean, like California's wildfires, have become fiercer, wilder, and less predictable, blind, sudden pummelings of rage.

I have become nervous, I realize, to write the word *tempest*, the word *hurricane* most of all, because these words have come to signify something so much worse than they did when I was growing up.

When I put pen to paper, I see Hurricane Maria again before me, the tempest that almost took my parents' lives, the whirling arms of wind flinging rooftops and bodies alike through the air, drunk, without tasting either, on blood and water, a blind cyclopean fury destroying the island I once called home. When I write this, I see a thing at once diablerie and dull nature, a monster made more monstrous by the chaotic shifting gears of our planet.

We live now on a planet in which once-in-a-lifetime storms happen over and over.

I do not know how to write about climate, I have learned, without writing about death.

I like lists.

Like long sentences, sometimes a list is the only way to record the *feel* of something.

To live in a place where the rules have changed, lists help. These are the things that have changed for me, climate-wise, since I was a child:

- In Dominica, our home had no AC. In New York, my current home, I do not know how to live without it
- Make jokes more often about the apocalypse
- Realize dimly that there might actually be an apocalypse
- Try to finish book before apocalypse, when publishers will still be around
- It is now normal (very normal) to feel the temperature shoot up or down, to see a blizzard just days after a balmy beginning to spring
- Consider being a mother and raising a child in places where, even in nature, you hear fewer insects, where you hear, instead, a blue absence that quietly unsettles you
- Consider raising a child and telling them about your favorite creatures that no longer exist except in zoos and legends and old movies
- Wonder how one gets hormone replacement therapy medication for transitioning during an apocalypse
- Realize you will be a woman who has avoided personal

apocalypse, even when the climate apocalypse comes, for you have come out, blossomed into the rude flower of yourself, and you would not be alive to consider such futures at all if you had not come out because the cramped pain of living in a closet was too much to bear and you almost attempted to kill yourself with poison and also you almost jumped in front a train because even after coming out your mother put you through so much guilt and pain that you felt hideous and alone and in need of help. Then, finally, you got that help. You realize only then that you have come close to taking Death's hand for that final ballroom dance too many times, and now, ultimately, you are okay. Even though the world is not okay, *you* are, sort of, now, and that is good.

- Despise politicians who still, like the forty-fifth president, think climate change is a hoax
- The hurricane shutters on our home in Dominica have to be bigger and thicker
- Start saving money now for all the trips, when you can still dive here or explore there or see this animal in the wild here
- Each hurricane season, think more seriously about what happens if family members die
- Love my partner and besties and future child more, and don't wait too long to let them know you love them

From childhood, my mother told me the signs that signaled the birth of great storms in the Atlantic. The clearest was the Sahara dust, a thin coating of desert brown that blanketed our veranda and outdoor chairs. It was a gift from the sirocco winds that blew, vast and enormous, through the deserts of West Africa and into the Mediterranean, then across the Atlantic.

The sirocco carries more than dust. Like all the great winds around the world—the *ghibli*, *xlokk*, *harmattan*, *sharav*, *simoon*, *föhn*, the Trade Winds, and many others—there is a long history of believing that the sirocco brings with it shifts in personality, gains in fortune, loss of fortune, a sense of sultriness and wildness, a sense of ennui and enervation, diseases, and madness. It carries life and death.

We expected those supernatural winds each year; hurricane season was simply an annual cycle, like autumn or winter in America's cooler states. Dominica had been hit in 1979 by a great tempest, Hurricane David, which left three-quarters of the island's population homeless; a reminder of that torrential storm still exists in the Botanic Gardens, where you can find a yellow bus crushed under a fallen tree. Dominica was also grazed by the outer bands of Hurricane Dean one year, the only hurricane I experienced while in Dominica.

Still, despite our preparations, the vast majority of hurricanes missed our island. For most of my life, I believed that the odds of a hurricane hitting us were low, never wondering if the very nature of those storms could change, so that they would hit harder, more often, more chaotically.

Even if the storms did not often hit us directly, we respected them, acknowledged their unsettling grandeur in the way Noah accepted the deluging vengeance of his deity.

A hurricane always reminded us, even before it arrived, that it had once been a god, the whirling, seething, howling deity with a single unseeing eye that, in its furies, the Caribs had called Huracan, and you knew it was divine, somehow, even centuries after the Caribs had named this unappeasable god of wind and sea, because you could simply *feel* it in the air, the dread and supplication and simply primal *fear* that preceded its arrival; the way the sky would sicken, then harden into a yellow-gray; the way the sugarcanes and grasses would be stilled; the way the raucousness of the birds and rustling lizards would disappear into an uncomfortable silence; the way the air was sharp with a quiet tension; the way the little villages lost in the mountains that I knew of only from the soca their radios blared on the winds would become tomb-quiet; and then the way the hair on the coconuts would rise and the palms would begin to bow and the lines at the gas stations would flare and everyone would yell at the frazzled attendants filling the tanks of cars, and the IGA supermarket would be packed as never before with people stocking up, and in other corners the old men who had been playing dominoes for half a century on the side of the road would scoff and say *what all you worrying about so, nuh, no storm coming*, and then the thunder would growl with the deep throat of some underworld hound and our power companies, already prone to outages, would warn us to unplug everything and it was usually too late because the lights were already flickering and we would pull in the furniture from outside in a frenzy and enclose our home in the darkness of our windows being shielded by hurricane shutters and we would find the hurricane lamps and candles and matches and flashlights and then just wait, wait, orbiting the stars of flames in our damp, dark homes, speaking until the fury of the rain began upon our roofs and drowned out all other sounds, and we would wait and hope for the best, knowing the hurricane would tug at the phone lines and would fling roofs and rocks in its razing rage and would uproot trees and would punch at our glass windows and feeble shutters, and if we were lucky, after it had left there would only be a few blown-off rooftops dotting the hills and fields, and only a few landslides blocking the roads, and just some branches and small trees on the grass, and soon we could return to our normal lives, with that deep, antediluvian *fear* faded to a dim pulse in the back of our heads.

Hurricanes, in other words, were a part of life—but from a distance. Even if the hurricanes always had the *chance* of hitting us, we generally had a sense of what to expect in terms of how severe and frequent the storms would be.

But by 2017, when Hurricane Maria destroyed Dominica, something had shifted. The hurricanes seemed fiercer, more unpredictable in their numbers and trajectories. The rhythms of the world had shifted, and, with them, our expectations.

I had not returned to Dominica since I came out. When my grandmother died, a sea away, I felt guilty about not feeling able to go to her funeral, even as my parents made it clear I was not to even think about traveling home, lest I cause them and myself ridicule or put us all in danger. Not long later, when Hurricane Maria ripped the island apart, I felt not only guilt over how far away I was, but also rage and terror.

Earlier in the afternoon before the storm hit, my father was still insisting, according to my mum's disapproving texts, that no storm was coming. My mother was frantically preparing as we always did for a leviathan storm, but my father, phlegmatic in his contrarian calmness, kept saying she was overreacting to the reports. We always worried too much, he said, and this storm would just veer away like they all did.

By nightfall, it was clear he was wrong. From New York, I watched, in the dull shock of someone not knowing what to do, as the weather radar sites showed a massive, fully formed hurricane go straight over Dominica. My texts stopped going through. On Twitter, I connected with other horrified Dominicans who could not reach their loved ones. I felt numb when I saw the prime minister himself announce, in a brief blip of news from the island, these chilling words: *Please tell the world that Dominica has been devastated . . . In the morning we will know how many dead there are.*

For days, we could not reach our family members, and I wondered, crying in my apartment, why I had not told my parents I loved them before our texts stopped, why I had assumed the storm would just pass. I wondered if my parents would die a sea away and whether my having come out had changed my fate so that I was not on the island to die with them. That the storm had my mother's first name only deepened my dread, because I could never say the hurricane's name without also saying her own.

Then, we reconnected. My mother told me that the island had been devastated beyond all belief. They had survived, but not without something like irony: they had only escaped alive, my mother told me, the queer daughter she had rejected, by hiding in a closet.

I blinked and chuckled, not knowing how to react to such inexplicable symbolism. To how close I'd come to losing them.

Later, amidst the wreckage, my father nearly died again. His health was failing. My half-siblings and I put out a call for the American soldiers who had gone to the island to search for him, and despite not answering us, they must have heard our pleas, though the soldiers later revealed that it was largely a fluke they found my parents' home at all, buried as it was amidst broken things. My father was medevacked to safety, first in Martinique, then in Florida, where he had multiple surgeries. As I write this, he is still sick, and my mother, always frail, has become a wraith.

The storm has passed, but it has not left them—and there are more storms still coming, still seething.

Storms like Maria never truly dissipate. They hit again and again, the trauma they have left in their wake reshaping our bodies and dreams.

(add to earlier list)

- Make memories you can cling to, and channel the strange way that, as Walcott said of the Caribbean itself in his Nobel lecture, love can put back together the fragments of a place or person after a disaster breaks them. Remember the love. A hurricane brings death; in its wake, somehow, there is also love, because it reminds us how close we were to losing the ones we care for. Live your life as much as you can, even as you fight with your choices, votes, speech, and actions, because it is easy to forget you can be deserving of love when the world is a slowly cracking vase
- Buy more panties
- Relax, sometimes, and remember that you deserve to

care for yourself, too, no matter how much the world has
broken around you

Earlier in 2019, I am in the Senegalese desert with my partner, a soft night wind tugging at our curls and clothes as we sneak off together to look up at the stars. The constellations are bright, sky-filling; behind us, a few dots of light glow from lanterns and torchlights near a row of large white tents that constitute the lodge we have decided on for the night. The tents have vanished in the black; all I can hear is the wind's low whistle.

We are trudging up a dune in Lompoul, holding a blue blanket and phones set to flashlight mode. The sand, earlier white-hot beneath my feet when we sand-boarded down the desert's frozen waves on an old white surfboard, has become a cool graininess. It clings to my feet. I pause two times to catch my breath as we ascend.

We have come here for a day and a night as part of a weeklong trip to Senegal to visit my partner's sister, who lives in Dakar. It is my first time in Africa, the world I have always felt both part of and distant from. It is woven, in blood-colored cords, into my ancestry through the transatlantic slave trade. It is where the soft brown of my skin and my tight black curls come from. West Africa, too, is the origin of some of my home's traditions, like *obeah*, the magic certain slaves brought with them and practiced in the Americas. Yet it is also a world that is not mine, for I have never set foot in it before now. It is familiar yet foreign, like this desert.

The night over the dunes of Lompoul is thick black, like oil, but when we look up it is like a great shawl, the wind rippling its stars. We lie on our blanket and stare at this *shahtoosh* of night. We hold hands, squeeze fingers, point out the majesty of the sky. We undress, begin to kiss under the stars, two bodies lost in the black. The air is cool but her breasts and chest are warm, then hot, and as she touches me I feel that strange thing they talk about in cheesy novels, where two become something that seems like one. We become one warm thing, fox fire, djinn smoke, queer girl magic in a country, like the one I came from, where queerness is not openly accepted. When we stop, breathing against each other and smiling and whispering that we love each other, we realize, again, that sand is everywhere, even in crevices it seems impossible for the sand to have reached. When we reach again for our pants, they release sand in little whooshes as we turn them upside down.

We walk back, wrapped in the night's dark too much to see each other, and quiet because we do not want a patrolman—if there are any—to find us, girls scandalous in love on the shipless waves of a desert. I cannot see her, except if I turn back with my flashlight, but I feel her pinky wrapped in mine. And I feel her smile, and, I hope, she feels mine.

It is like coming back from a brief dream, a moment in time when we got to escape from everything and just *be*, under the Milky Way. A moment when the climate of a world felt perfect. To make this trip work, we have largely had to pretend to be friends rather than partners, only touching each other when we are at her sister's home or away from anyone else's eyes. I hated that; I felt distant from someone I loved, even when I was sitting beside her. Here, under the stars, it's different. Here, we got to feel open, indulgent—a time to be alone together, and to breathe deeply, and, for at least a moment, forget the horrors, and just exist in each other's embrace.

There are many moments I remember on this trip, like the sudden rise you feel when you sit on the back of a camel; the funny beauty of hearing tinny recordings each morning from a mosque as regular as cockcrow; a man who carries around birds in a cage and releases them if tourists pay; the way many signs are in the colonizer's language, French, rather than in Wolof, the language most Senegalese people speak, because that is what colonialism does to a place.

But it is this moment, under the stars, when we got to vanish and burn bright all at once for a bit, that I remember most.

I think often of the small and Brobdingnagian ways my world has changed, but no matter what, I still want to live my life and feel love and be happy.

I still want to ride the wild seas like a mermaid, want to smile in my warm moments of solitude when I can rely on things being still and calm, want to dream of walking on the bottom of the ocean and the top of the moon alike.

I still want to sail the deserts and walk on the dunes of the sea.

I still want the stelliferous beauty of dreaming of what the night sky can hold.

I still want to settle down somewhere and be the woman I have always wanted to be, in love with another woman, and helping to raise a child.

I still want to dream of all the things the color blue can mean, without fearing these dreams are luxuries in a slow-burning, slow-erupting, slow-cracking world.

I want to still believe, as I have for decades, that I am just a collection of shifting particles on a planet that came into existence by chance and physics, and not because of some theistic determinism. I want to believe that, even if there is no grand meaning for our lives and our planet has a finite lifespan—as do our art and dreams—that art is worth making and love is worth finding. That it's worth fighting to preserve a world where dreams are still possible. I want to be able to sit for a moment and forget the Lovecraftian horrors of planetary devastation, and just enjoy life and love, for a bit, because caring for ourselves, others, and our planet goes together.

I want to remember what Audre Lorde wrote in "Man Child," her seminal 1979 essay on being a Black lesbian mother, where she reflected that "when I envision the future, I think of the world I crave for my daughters and sons. It is thinking for survival of the species—thinking for life." I want my daughter or son or beautiful binary-defying child to have the world I did. A better world than I did.

Because without those dreams, I do not know how to live. I want to help keep this world going as long as I can, so that the next generation can live in it, too, and so they, too, can face fear and sea-dream and find love under the night's shifting shawl of stars.

A BRIEF HISTORY OF BREATHING

PITCHAYA SUDBANTHAD

Whenever the air in Bangkok becomes too polluted to breathe, my relatives turn to the internet. They open an online city map to gawk at air-quality measurements, the areas lit up in colorful shades—green for the clean zones and yellow to indicate moderate pollution, interspersed with patches of unhealthy orange and hazardous purple. They click on markers that zoom in on individual neighborhoods, which give hourly readings from the past week and forecasts for the next, to see when they might be able to breathe without worry again.

These readings come from web-enabled air-quality monitors sold by an air purifier company. My relatives often debate whether the numbers have been inflated to help the company sell more of these units. At the same time, there's also talk that the government's official particle counts have been reduced to keep the populace from becoming too alarmed. One of my aunts carries around a mini air-quality detector that she uses to measure particulates in her home and those near the busy main road, just to take note of the difference. These are the numbers she trusts most, because she can see them, and breathe them, for herself.

It's not hard for me to tell when there are too many particles adrift unseen. My nose gets itchy and runny. The air holds hints of burnt rubber and over-boiled eggs. On busy roads especially, the smells seem unavoidable. Many people, from cops to motorcyclists to street hawkers to tourists donning "I love Bangkok" T-shirts, wear disposable surgical masks over their mouths and noses. Teenagers often sport ones printed with anime figures or bright, colorful designs. If one has to suffer from breathing in poisoned air, one might as well do it with style.

In Bangkok, particulates that are 2.5 micrometers and smaller, or PM2.5, are a topic of small talk as favored as the weather, traffic, and football league results. There's good reason to keep an eye on those numbers: PM2.5 particles can settle deeply into lungs. Every year, tens of thousands of Thais prematurely expire from their effects. To talk of the air is to convey one's concern about matters of life and death. When the readings are bad, we remind each other to wear filtering face masks and pray to dead ancestors for health. When they're good, we thank the higher entities for rain and wind, and breathe in the air to the fullest capacity of our lungs, trusting that the day's color code of green is accurate.

I don't remember anyone wearing masks in public when I was a teenager. It wasn't that Bangkok air was so much cleaner back then, but unless I ventured to walk along one of the city's traffic arteries during rush hour, I didn't have to fear the ethereal or invisible, beyond the norm of ghosts and bad karma. Since the early 1990s, the population of Bangkok has tripled as more people have arrived from far-flung provinces in search of better economic opportunities. The shadows of skyscrapers overtook shophouses. Housing developments replaced green rice fields and marshy grassland. Dust from the construction boom and ash from trash burned in growing neighborhoods mixed with dust from new roads built to accommodate the explosion in traffic. The number of cars and motorcycles in the city grew to the single-digit millions, then to double-digit millions, on streets and highways unequipped to handle the volume. Of course, it is not just dust that comes out of tailpipes. They also expel exhaust fumes filled with carbon gases.

As global lust for economic growth worsens the air quality in cities and towns around the world, more carbon dioxide is also being spewed into the greater atmosphere. Climate change contributes to drastically altered weather patterns, and those in turn regularly make the air in Bangkok more stagnant, while hazardous ozone and particulates permeate the air at ground level and stay there. In places where we now see far drier weather, more frequent and intensified fires are consuming entire landscapes—like those of the past few years in the Amazon, Australia, and Indonesia—and sending plumes across thousands of kilometers. The burning releases even more carbon gases and poisonous particulates, continuing the entwined loop of climate catastrophe and our struggling respiration.

I often confront my own complicity as I am fed into the slow, perpetual circulation of Bangkok traffic. Even with the addition of Skytrains and a subway system, traffic jams haven't improved. I loathe having to go anywhere in a car during rush hour, but the trains convenient to me reach only the major commercial thoroughfares and not my destination. So I often sit in a car, sometimes for hours, just to travel short distances, and watch exhaust fumes rise from the tailpipes of idle vehicles, thickening the air. If I'm on an elevated highway near sunset, I can marvel at the orange-hued haze that makes the city look like a shimmering mirage in the desert.

I could smell the still-smoldering towers whenever I stepped out of my apartment building in Downtown Manhattan. Sometimes, I could even see a haze rising across the West Side Highway, where workers were disassembling the crumbled, hulking ruins and recovering the physical remnants of extinguished lives. Just over two months before, I had watched the towers aflame from an office building ten blocks north, and I was still in disbelief that I had been witness to the sudden deaths of thousands. Reminders lingered everywhere—in missing-persons flyers still taped to walls and light poles, in the near-constant siren screams of emergency vehicles passing by stilled traffic on Greenwich Avenue.

A week after the attacks, I went back to my apartment building for the first time. I strapped on a dust-filtering mask and boarded a truck along with other area residents. It took us from a FEMA center set up on Houston Street and down the riverfront, past military checkpoints, and into the blocked-off zone that had been deemed too dangerous for prolonged human exposure. We drove toward Battery Park City through the scattering of wreckage and debris, its violence made stranger by the brightness of an otherwise normal, sunny city day. I expected the worst. I had watched—from the SoHo block I'd reached on foot that day—the remaining North Tower crumble floor by floor, no more than a block from where I lived. I was surprised to see my building still intact. When the towers disintegrated, I thought they had crushed my building with the weight of their fall.

Inside, my apartment was covered almost entirely in a thin layer of gray. The dust cloud had blown through the half-inch of window I'd left open before departing for work that day, as well as a quarter-sized hole that had later been punched through the glass by an unknown projectile. I imagined the gray matter to be pulverized bits of planes, buildings, and people. I paused to consider whether a sacred ritual was required before touching it, but then just mopped and wiped it away.

Tens of thousands of liters of volatile jet fuel had conflagrated to heat the steel skeleton of the towers to the point of structural failure. Its particulate remnants turned into a foul gray plume that floated over Lower Manhattan before the wind blew it southward to the ocean. To help allay fears of asbestos and other hazardous material in the air, the authorities gave away HEPA air-filter units to building residents. When I turned on its ionizer for the night, the air started to smell of its metallic charge. The next morning, I woke up to find previously buoyant particles settled on the pair of glasses I'd removed before bed. A couple of months after the attacks, the air outside still smelled strongly of burning. The news reported that first responders at Ground Zero were experiencing difficulty breathing. Glass fibers, dioxins, and who knew what else mingled invisibly in the air. I bought N95 respirator masks at the hardware store for my walks to the subway or grocery store.

Soon, what was left of the dust was being washed away by rain. Winter was arriving. Snow would soon cover the ashes settled into the crevices of the charred, twisted steel ruins, which still awaited dismantling. Every once in a while, I would hear the sounds of demolition explosions, preceded by the screams of air horns. But slowly, life began to return to some semblance of normal. The air became more breathable. Instead of respirator masks, I only covered my mouth and nose with my winter scarf, the way Bangkok motorcyclists often wear bandanas.

Over the next year, the Bush administration would use false information to justify America's bombing of and eventual war in Iraq; there would be no repercussions to the native country of the 9/11 hijackers, a major military ally and oil supplier. Frustrated by the denial of truth and the promise of more senseless, baseless violence, I joined the hundreds of thousands gathered in Union Square to rally against what was clearly a war for fossil fuel and little else. Someone gave me a button. On it was the famous Milton Glaser "I love NY" logo with the addition of "and Baghdad." For a second, I thought it said, "and Bangkok."

I was made to wonder: did the tourists wearing the "I love Bangkok" T-shirts truly love the city where my family lived and breathed, any more than American policy hawks loved the notion of egalitarian democracy in the Middle East? Technology's collapsing of distance has rendered every region a place where some variety of value can be extracted, be it exoticized leisure or fossil fuel. When people laud a new world made smaller by air travel and the internet, they often neglect to mention that these new intimacies have been made possible only by ravenous empire.

After the protest, I returned to Battery Park City and took a walk along the promenade by the Hudson River. The sun was beginning to set behind the Statue of Liberty. I gazed at the cargo ships and oil tankers in New York Harbor floating out to the Atlantic Ocean, and perhaps somewhere beyond.

In the early 1980s, the beach along the highway between Al Khobar and Dammam was popular with the immigrant workforce that came to Saudi Arabia in search of better pay, a group that included my father. When we went there to picnic with friends from the architecture firm where my father worked, I was a Thai kid surrounded by Indian, Filipino, and Korean families, or more often, groups of young, able-bodied men from everywhere in the world, on a break from their jobs at oil-field service companies.

A reliable breeze blew in from the Persian Gulf. In the high desert temperatures, the wind and seawater were always welcome. My home city of Bangkok was no temperate place, but the heat in Saudi Arabia felt different. The sun was quick to scorch exposed skin. As a child in Bangkok, I couldn't care less about sunscreen. In Saudi Arabia, I begged for it. It didn't surprise me how much of the landscape was lifeless. I observed it every day, looking out the window as I rode to school on a long, thin highway: expanses of dunes on one side, the sea on the other.

At the beach, I played with the abundant sand, digging it up to build what my megalomaniacal seven-year-old mind thought to be a city—like New York, which I'd visited for the first time that year. The pits I dug widened as my sand cities rose. If I dug far and long enough, I might reach the oil deposits underneath.

Long ago, minuscule sea animals had died and settled into the sunless, airless depths of a long-departed sea. The concentration of their biological sludge and high temperatures turned them into crude oil sealed by years of sediment deposits. In 1938, the Standard Oil Company of California found those deposits, and since then, millions of barrels a day had been pumped out of the ground. Some of that oil was hanging from a key chain that I carried with me in Saudi Arabia. Through a dark glass vial, I could see an inky liquid roll as I tilted it.

The empire of capital would not let the dead, ancient animals alone. Through systemized extraction, they were being unearthed to light up and power our cars, motorcycles, and sky-shattering jets, and they were vengeful at us for disturbing their rest. In their incineration, the dead would turn into dangerous dust, seeing to it that our lungs became poisoned and diseased, that the entire world grew hotter, that seas flowed over our cities of stone, and that lush forests dried into desiccated dunes. The dead want to quicken our union with them, perhaps so that we may sooner know what it is like to be exhumed for some living being's expedient use. In the end, we become all that we consume.

Where I stood then—sandy toes wetted by ebbing waves—I was too young to see that the tankers floating toward the horizon carried both the dead and the future. I could not yet understand that this place, where the desert met the sea, was inseparable from an American metropolis overlooking a busy harbor and the growing city where I was born.

As afternoon approached twilight and the tides crept closer, I got up from my sand city. The waves would have it. On the walk back to my parents' car, I picked up a sun-whitened seashell, its inhabitant long taken by the deep. Hello, friend, I said to it, and then I breathed.

WHAT WE DON'T TALK ABOUT WHEN WE TALK ABOUT ANTARCTICA

ELIZABETH RUSH

Sometime last spring you receive a cryptic missive from your program officer at the National Science Foundation. It reads: *An interesting opportunity has come up. Call me in the morning. Valentine.* A strong wind blows all night, stripping the cherry blossoms from the trees. Your husband steals the covers, and you, tucking the comforter under your right leg and turning over, slowly steal some of them back. After your shower you sit down to write and only then do you realize you missed a call.

When you return it, the phone rings not even one full time.

"Hello, Elizabeth," Valentine says.

You think it is either very good or very bad that she picked up so quickly.

Valentine tells you that she spent the previous day in a planning meeting for the International Thwaites Glacier Collaboration, a five-year program to study what is one of Antarctica's most important and least understood rivers of ice. "This year they're deploying an icebreaker to investigate. There's one berth remaining and I recommended it be given to you," Valentine says. Your heart rate rockets. Thwaites is one of the few glaciers you know by name. If Antarctica is going to lose a lot of ice this century, it is going to come from Thwaites. That's because the glacier rests below sea level, making it vulnerable to warm water incursions that cause rapid melting from beneath, which, in turn, could force the entire glacial system into accelerated collapse.

Collapse is a word you hear often these days: as in Antarctica's second-largest colony of emperor penguins recently collapsed and a recent United Nations report that says that "a million extinctions and ecological collapse" are on the way. In response, environmental activists around the world stage "die-ins," a kind of political theater they hope illustrates how our actions have the power to both make and unmake the world of which we are a part. Meanwhile, Wall Street is betting that the president won't let stocks collapse, and a Brazilian mining dam collapses, unleashing a tidal wave of toxic mud, killing more than 240 people. British Steel and Venezuela are both "on the verge of collapse." Together we teeter on the knife-edge of unprecedented change. And you, lucky you, collapse into your chair to talk to your program officer about your upcoming trip to Antarctica.

Thwaites alone contains three feet of potential sea level rise, and were it to wholly disintegrate, it could destabilize the entirety of the West Antarctic Ice Sheet, causing global sea levels to jump four times higher. In terms of the fate of our coastal communities, this particular glacier is the biggest wild card, the largest known unknown, the pile of coins that could tip the scales one way or another. Will Miami exist in one hundred years? Thwaites is the answer. At least that is what many scientists think, which is also why *Rolling Stone* started calling it the Doomsday Glacier a couple years back. But no one has ever before been to Thwaites's calving edge—no one knows just how rapidly the underside of the ice is melting—so many of the ideas we have about how this glacier will behave are a mixture of science and speculation, meticulous modeling married to our increasing fear.

It is late spring in New England and the leaves on the oak trees are a tart, electric green. Classes have ended. You walk to the library with a watchful heart, steeped in the sense of setting out. Snake up the three flights of stairs to aisle 63A. First impressions: scant. The total number of books on Antarctica in the "History, Geography, Travel" section of the Brown University collection can't total more than fifty. In order to begin to sink into this place through the experience of other people, the pool you will pull from is relatively small. Most of the books have foreboding names like *The Worst Journey in the World*, *Deep Freeze*, and *Where the Earth Ends.*

Back in your office, as you stack and inspect the nearly two dozen tomes you withdrew, you realize that only two of your selections were written by women. The introduction to Sara Wheeler's *Terra Incognita* confirms your hunch. She writes, "Men had been quarreling over Antarctica since it emerged from the southern mists, perceiving it as another trophy, a particularly meaty beast to be clubbed to death outside the cave." You laugh out loud to no one but yourself. Think that her book has promise as you set it aside to flip open the second, penned by someone with whom you, at least, share the barest of affinities. You laugh again, but this time for a whole different set of reasons. Like half of everything ever written about Antarctica, Caroline Alexander's *The Endurance* is, not surprisingly, a history of the famous failed Shackleton expedition. You should have known, since his boat that sank in the Weddell Sea carries the same name as the book that lies before you now.

During the majority of the two hundred years since humans first saw the southernmost continent, women weren't welcome. Rear Admiral James Reedy, the commander of the United States Naval Support Force in Antarctica in the 1960s, referred to it as "the womanless white continent of peace." When the *New York Times* journalist Walter Sullivan wrote of the first all-women scientific expedition to the South Pole later that decade, he described the undertaking as "an incursion of females" into "the largest male sanctuary remaining on this planet." It wasn't until 1974 that Alice McWhinnie, the first woman to head an Antarctic research station, wintered-over there with her required "assistant," a biologist and nun named Sister Mary Odile Cahoon. In the intervening decades, many women who made it on board the ships and out to the research stations experienced harassment and assault while working in such a remote location. When one friend heard about your Antarctic expedition, she suggested taking personal defense classes; another wanted to know just how many other women would be with you on the boat. And a third sent you an article about a Boston University professor who was being investigated for taunting and degrading his female Ph.D. students alongside the glacier that carries his name.

All summer long you submerge yourself in the Antarctic

canon and the desires of those who journeyed south to conquer a thin sliver of the continent of ice. Some books you cannot finish. Others leave you pleasantly surprised. By July you are growing bored. The same half dozen events—Scott's death eleven miles from One Ton Depot, Shackleton's miraculous return, Mawson shooting and eating his sled dogs—are woven into nearly every narrative account of the last continent's history. Then one painfully humid day you read Ursula Le Guin's short story "Sur" and enter an entirely new Antarctica.

In Spanish, *sur* means "south." *South* as in the singular goal of the "Heroic Age of Antarctic Exploration," the quest to reach the southernmost land—a desire undeniably driven and largely dominated by countries from the global north for whom such folly was affordable. *South* as in Sir Ernest Shackleton's personal account of the doomed voyage that would turn him into a legend. In Le Guin's deft hands, the word and the world it describes take a radical turn. Her female Latin American narrator begins, "Although I have no intention of publishing this report, I think it would be nice if a grandchild of mine, or somebody's grandchild, happened to find it someday; so I shall keep it in a leather trunk in the attic, along with Rosita's christening dress and Juanito's silver rattle and my wedding shoes and finneskos."

You stop reading and google *finnesko. Finnesko*: from Norwegian, a soft-hide boot of tanned reindeer skin used for cold-climate travel. As in what this woman wore on her feet when she went to Antarctica. *Finnesko*, a relic of exploration that now, in some mind-boggling turn, appears alongside a bunch of domestic baubles. Sterling and animal hide, nested together in a chest on the top floor of a small suburban house in Lima. You feel the boundaries long built between two gendered ways of inhabiting the world blur a little. It is exhilarating and you're only on the first sentence.

Our narrator, who does not have a name and therefore could be any of us, gathers with nine other women—one Peruvian, three Argentines, and four Chileans—in Punta Arenas in 1909. Together they cross the great Southern Ocean in a little steamship they call *la vaca valiente* (or the valiant cow); they raise a glass of Veuve Clicquot when they see their first iceberg, and another when they finally spot land. They encounter Scott's famous Discovery Hut, inside which tins of tea have been left open, the floor scattered with stale biscuits—a polar bachelor pad. Teresa suggests cleaning it up and using it as their camp; Zoe wants to set it on fire. They do neither. Berta and Eva build an ice cave instead and name the central chamber Buenos Aires. In that cold home the women play banjo, carve ice sculptures, and plan their expedition to the pole. After months of preparation pass with the placing of supply depots and physical acclimation to the cold and the wind, the narrator and five others set out, dragging heavy sledges into the great silence, hoping to reach the southernmost spot on Earth by foot.

Like any other early explorer, they name, but in this case "not very seriously," the land features they encounter. What Shackleton calls the Beardmore is briefly known as Florence Nightingale Glacier. Just south of it lies Mt. Bolivar's Big Nose. On December 22, 1909, they become the first people to ever reach the South Pole. The women talk about leaving some kind of mark or monument, a flag perhaps, but decide there is no point. "Achievement is smaller than men think," Le Guin writes.

"What is large is the sky, the earth, the sea, and the soul." You underline these lines. Instead of snapping photos to commemorate their accomplishment, the ladies drink a cup of hot tea and turn around.

The real climax comes later when, upon returning to the ice cave, the sledging party discovers that Teresa is pregnant. She goes into labor on the ice and screams herself "hoarse as a skua." The women of the expedition tend to her, for many have labored themselves. After twenty long hours little Rosa del Sur is born and the women drink their last two bottles of champagne toasting to her. For the first time since you started researching the great southlands you sense that the story you just finished does not end when the text runs out. Instead of consecrating the closing of a possibility—anytime anyone is the first person to do something means that no one can claim that title in the future—"Sur" celebrates the everyday act of giving birth and the host of unknowns and possibilities that attend the arrival of a new life. Long after you stop reading this, the story continues to unfold.

For the majority of human history no one glimpsed, let alone set foot on, Antarctica. There is no indigenous history to reference, no deep institutional knowledge of the ice. This is part of the reason why when we speak about that faraway continent we almost always talk in firsts. The first to cross the Antarctic Circle: Captain James Cook in 1773. The first to spend an entire year there: those aboard the RV *Belgica* in 1898. Roald Amundsen is the first person to reach the South Pole in 1911, narrowly beating out Robert Falcon Scott, who will die in a tent some months later, just a few miles short of the food cache that would have saved him and the remaining members of his team from starvation. Their deaths—some say orchestrated by Scott in order to turn second place into something tragic and noble—along with the many others, racked up by those who sought the extraordinary privilege of marking our map's largest blank space, will haunt a surprising number of the stories we tell about this place that we only recently began to know.

Just a month before setting sail, you read about two men—Colin O'Brady and Louis Rudd—who, according to *The New York Times*, "hope to conquer a continent that has become the new Everest for extreme athletes." When you encounter this line, you will not lie, you are furious. So much time had passed since the initial exploring expeditions that remain so famed in Antarctic history—the planet deep now in the throes of the single largest geologic transformation human beings have ever witnessed—and yet mainstream journalism's representation of Antarctica appeared, at least in this instance, not to have evolved a lick. For many, Antarctica remains a barren and brutal faraway place, unconnected to human civilization; as Sara Wheeler writes, "the continent was little more than a testing ground for men with frozen beards to see how dead they could get." To pit O'Brady and Rudd against the elements is to continue to drive a wedge between man and nature, to make this place that we know so very little about into a hurdle, a hindrance, a problem. This language turns the more-than-human world into a placeholder, a poetic metaphor, a thing that must be overcome in order for the singular human being to rise up. Well, you think to yourself, I don't buy it. The longer we talk about Antarctica as a place to be conquered, the more we steal from ourselves the possibility that its transformations—the ones we have set in motion with our insatiable appetite for more—are, whether we like it or not, in the process of transforming us.

You'll be the first to admit that gender essentialism comes up again and again when you're thinking about Antarctica. As though the presence of womanhood infuses a sense of nurture into the Antarctic night, making the human communities on the last continent magically less competitive, warmer, and more welcoming. Which, of course, reaches even deeper into our cultural baggage, unearthing the old idea that women are made for mothering, and not for adventure or intellectual pursuits. It is this kind of reductive thinking that is at the root of what kept women at arm's length from Antarctica for much of the last two hundred years. But just as you are skeptical of making too much of an argument that sees women as inherently different than men, it is also equally hard to imagine that the sudden arrival of those whom our society calls women made no discernible impact on how we understand the only continent on Earth so harsh it lacks indigenous inhabitants.

Eventually you come across an article by Mark Carey, professor of history at the University of Oregon, that investigates the interplay between gender, glaciers, and the science employed to better understand the latter. He writes, "Scientific studies themselves can be gendered, especially when credibility is attributed to research produced through typically masculine activities or manly characteristics, such as heroism, risk, conquests, strength, self-sufficiency, and exploration." The characteristics he lists not only describe the majority of the tales you read about Antarctica in preparation to deploy to Thwaites; they also, in sly and determinate ways, continue to define many of the stories we tell about Antarctica, even today. And as Carey points out, that not only impacts the narrative arc we build around the continent of ice, it also goes so far as to shape the science that takes place there, valuing the triumph of individuals over the elements at the cost of recognizing how inextricably entwined those researchers are with their support networks and the more-than-human world of which they are a part.

On Amazon you discover that of the top fifty best-selling books about polar regions, not one is written by a woman about a woman's experience in Antarctica. Of the top one hundred, not a single female writer of color appears. With that, you decide to devote the rest of the month prior to your departure to reading Antarctic writing by women. You dive into books like *South Pole Station*, Ashley Shelby's fabulous (if fictional) account of how having a climate-change denier on the ice tests the weave of the southernmost human outpost on Earth, and Gretchen Legler's *On the Ice: An Intimate Portrait of Life at McMurdo Station*, in which she writes about her love of Edward Wilson's watercolors and how she met her future wife in Antarctica.

You had heard about Dr. Jerri Nielsen's *Ice Bound* about a year prior when you were talking with one of your writing students about applying to the National Science Foundation's Antarctic Artists and Writers Program. His mother had listened to it on tape back when he was in grade school. "It's about a female doctor who spends the winter at South Pole Station and contracts breast cancer," he had said. Then his voice dropped to a whisper. "I think she had to cut off her breast herself to survive."

Great, you think, women who go to Antarctica must remove mammary glands in order to not die. Now, that's a message I can *really* get behind.

As it would turn out, what your student remembers is only half true. Nielsen did indeed contract breast cancer while stationed at the bottom of the planet. Though she does not perform a mastectomy, but rather a biopsy and eventually, after chemotherapy drugs are dropped from the belly of a plane flying over the pole, she administers them herself until she can be evacuated some months later. Her cancer will go into remission long enough for her to write the story of the ordeal. What stays with you long after you finish reading is not the nitty-gritty details of the doctor's medical condition but the way she describes the community that forms around her in the middle of the long Antarctic night; it is a community of care and close attention paid to all of its members, especially to those who ail most.

At one point she writes in a letter to her family about why she and the others are at the South Pole in the first place. It was not conquest or science, she argues, that drew them down to Antarctica. She writes, "'We' are why we are here. We are here for each other. The longer that we live together, the more love and respect I have for everyone. People I wouldn't talk to in the world, I relish seeing in this place. We come to understand and rely on each other in a way that is not of this century, not of this time. This is how human beings were meant to live."

It's not that comradery isn't part of Shackleton's or Scott's books; it's just that when it does arrive it serves as a set piece, the backdrop against which tales of derring-do and exceptionalism take place. But in *Ice Bound*, as with the majority of the other books you read by women about Antarctica, the act of coming together is the accomplishment itself. The firsts have all already been tracked down, clubbed to death, and left outside the cave door. What remains is what the ice demands: that we work together to survive, nothing more, nothing less.

Nearly a year later, after you spend two months on a ship sailing to Antarctica with fifty-seven other people (sixteen of whom are women); after the medical evacuation of one of your shipmates and the ping-pong tournament, after you make it to Thwaites and after you watch a 375-square-mile chunk of it break apart into hundreds of pieces, after sea levels rise a little higher, after the scientists extract their sediment cores from the sea floor, and after they place remote sensor tags on elephant and Weddell seals, you return home. Back in Providence, it is not the ice or the adventure you miss, but the people you lived alongside and, in particular, the women of your expedition with whom you shared a kind of secret: that working together in Antarctica intensified the weave of your little community, to the point where, in those select and shining moments when simply getting the job done was all that mattered, nothing made any one person different from the next. You were hands and hearts and minds gathering what information you could about what has happened in the planet's past so that you might better understand the future we share.

Upon return you discover a new addition to the small canon of books by women about the southernmost point: Elizabeth Bradfield's *Toward Antarctica*. You like the thin volume even before you crack its spine because the title suggests that the goal is always shifting and what matters is the journey you take toward any kind of knowledge. Bradfield's book is humble. It is not a record of her being the first person to do anything. Instead, it is a laborer's tale, full of the kind of language those who work as support staff on ocean-based expeditions to Antarctica use. It is replete with words like *bulkhead* and *Zodiac*, *float coat* and *full landing gear*; and it recounts in sharp lyric prose the ways in which a ship is a metaphor for a republic or even the entire Earth. She closes her book with a question, "But then, one fragile hull holds us all, doesn't it?"

If your time in Antarctica taught you anything it is that the most important thing you can do as an individual is to recognize that you are not an individual but part of something larger than yourself, be it the group of people on board the *Palmer* who during your two months at sea became your family, or those whose voices your own joins with to sing your protest chants and pronounce that you too are part of a world where the climate is changing and so can we, where Antarctica is transforming and so must we. Just as individual resilience means little on the coldest, windiest, driest place on Earth, individual actions alone are incapable of addressing the comprehensive way in which the climate crisis is remaking the planet. From the women who went before you and with those whom you lived alongside, and perhaps, most profoundly, from the continent itself you learned this much: collective action alone will keep us alive.

IOWA BESTIARY

MELISSA FEBOS

> And I've never found a way to say I love you,
> But if the chance came by, oh I, I would
> —DAR WILLIAMS, "Iowa (Traveling III)"

1.

As I drive over the hills of Iowa for the first time, Donika reads me a list of the animals that inhabit our new home. I am struck by their overall mildness—no bears or wolves or poisonous snakes. Maybe a bobcat or a coyote—much smaller than the coywolves of the Northeast—but nothing fierce enough to pose any threat to, say, a five-foot-tall woman on a jog through the woods. *Are there woods in Iowa?* I ask her to look it up.

Early in our consideration of a move to the Midwest, my brother, who works in sustainable technology, told us it was probably the most pragmatic place to relocate to in light of sea level rise. I have lived in the coastal Northeast for my entire life and spent most of my adult years joking ruefully about how those of us at sea level were going under any day now. It mostly stopped being funny in 2012, when Superstorm Sandy flooded south Brooklyn, a mile or two from where I lived, and ravaged the Jersey Shore, where I worked, rendering many of my college students homeless. Maybe it should never have been funny at all, but humor is one way to cope with powerlessness.

I sometimes think in geologic time when I get scared, try to forget the anthropocentric view entirely. I envision a time-lapse reel of the post-human Earth, its terrain undulating with millions of years' worth of natural revitalization and annihilation, the framework of human values refreshingly obsolete. If I zoom out far enough, the tragedy wrought by human life shrinks to a wretched moment, or simply a moment of transformation, like the Pan-African orogeny that formed supercontinents some 600 million years ago, or the Cretaceous-Paleogene extinction event, which wiped out three-quarters of Earth's plant and animal life 66 million years ago. When I find a frame in which neither I nor anyone else seems to matter, the powerlessness I feel hurts less.

I wept as we drove out of New York City on the Brooklyn-Queens Expressway. For months, I'd feared the pain of leaving. Call it heartbreak, if that is what follows the choice to leave something you still love. The city has been fused with my self-conception for my entire adult life. However ready I may feel and though I know the reasons, departing is a process of prying myself apart—more cleaving than leaving.

As we exited the metro area, through the smelly industrial parts of New Jersey, into the lush mountains of Pennsylvania, my grief evaporated, though I know it isn't gone for good. Grief isn't always possible for me to feel while still in the process of losing something. Sometimes it arrives too late to say goodbye, once I am already gone, or they are.

We arrive in Iowa in early July and the temperature and humidity seem to rise daily. It is hotter here than Brooklyn in the summer, but the heat is softer somehow, without all the glass and concrete—like convection taking place inside a mouth instead of an oven. I enjoy being wrecked by the heat, which makes it harder to think, or to feel anything else.

Our first week here, I sit in the backyard, listening to the leaves, waiting to feel sad but not feeling much of anything but the quiet, which is draped like a soft green blanket over everything. I inhale the smell of outside—trees and dirt—and watch a mosquito hover around me. She is huge and lethargic, hanging in the air as if from theater wires, lurching toward the smell of my blood, which must babble to her like the nearby creek—where she was probably born—does to me.

When the mosquitoes multiply, I go inside and read about a study out of Iowa State that has found a new species of mosquito, *Aedes japonicus japonicus*, likely drawn to the area by the rising temperatures and increased rainfall.[1] An invasive species, they are also robust vectors of arboviruses—the kind carried by insects—and have been known to transmit West Nile and multiple kinds of encephalitis.

Climate change helps some species to flourish. As vectors like *Aedes japonicus japonicus* and ticks thrive in warmer temperatures, so do diseases like Lyme and West Nile. When exposed to higher levels of carbon dioxide, even poison ivy surges in size and breadth and produces a fiercer strain of its rashy chemical.

And yet, arboviruses scare me a lot less than those spread by humans.

In the months before we left Brooklyn, rubber gloves and paper masks littered the sidewalks. Ambulance sirens rang all day and night. The windows of our favorite restaurants wore hasty plywood shields with spray-painted crests. At sundown each night, fireworks exploded in the alleys of our neighborhood and sometimes continued until dawn. Our dog shook under the bed in which we slept a fragile sleep and awoke brittle with exhaustion. A scaly rash emerged on my hands from frequent washing, and stress hives bloomed on my chest and neck. We weren't going because of the pandemic, but sometimes it felt that way. So little of our life there remained.

When we left New York, my sorrow mingled with relief to be leaving the worst of the virus behind, despite all the predictions that it would follow us to the Midwest, which, of course, came true.

2.

I am thriving here. I go for long runs outside and roam the massive grocery stores, drunk on the bounty after twenty years in the narrow aisles of Brooklyn bodegas. Each week, when I return to the midwestern supermarket, it has the same things in stock that I bought the previous week. The reliability lulls me into attachment, creates the illusion of other things staying unchanged. I eat the exact same foods every day, every week, with the gusto of the previously deprived.

One morning around six, on my favorite jogging route along

the trail beside the Iowa River, I cross one of the footbridges and see a hummingbird. I think of that Mary Oliver poem "Long Afternoon at the Edge of Little Sister Pond," which reads:

> dog love, water love, little-serpent love,
> sunburst love, or love for that smallest of birds
> flying among the scarlet flowers.

It is, of course, a poem about death.

This hummingbird whirs around the bars of the fence that lines the footbridge, hovering at the corner of the interstices. I assume that she is drinking water, as it rained in the night and glimmering drops still cling to everything. When I get home, the internet tells me that she was more likely collecting bits of spider silk to build her nest. After glomming them all over her beak and chest, she'd have returned to her nest-in-progress, where, combined with lichen, feathers, moss, and fur, the silk would provide an adhesive, like duct tape, to cohere the tight little cup in which she'd hatch and tend to her offspring.

The rising summer temperatures here, and everywhere, are a problem for hummingbirds, who are reluctant to search for food (and subsequently pollinate) and to mate when it gets too hot.[2] If the nights are cool enough, their body temperatures sink by half and they slow into torpor to conserve energy and recover. But these balmy nights limit the energy their small forms can save.

Just as rising temperatures cause plants to bloom earlier ir spring, they narrow the window between the hummingbirds' win ter respite and their return. Scientists estimate that within two d cades, hummingbirds will miss the first blooms of spring entirel

Some days, I am in a kind of torpor, too. While my body moves almost constantly—unpacking boxes, running alongside the Iowa River for miles and miles, calming only at night or when I stand outside, looking—my thoughts thicken like hot cereal as it cools; my emotions indecipherable, as if being in a strange place has made me a stranger to myself.

Still, I try to notice more. Self-recognition is not the only form of recognition. Perhaps it isn't even the most important.

3.

Spider silk hangs from everything here, gauzily haunting doorways and tree branches to remind us of careful work undone, of lives washed away in the ferocious storms of this region.

For a whole month after we arrive, we watch an orb weaver build a web between the decorative metal pillars of our small porch. A member of the spider family Araneidae, *Larinioides sclopetarius* is a beefy but elegant spider with striped legs who prefers to build on steel, and is thus sometimes called a "bridge spider."

Once, when I was eight, after I had called my father into my room to kill a spider, he explained to me in his effective but not always comforting didacticism that, if not for spiders, we would all be eaten to death by mosquitoes. Until then, it was as if I'd never really *seen* a spider. I haven't intentionally killed one since. It wasn't the fear of a terrible death that changed my relationship to them. I nurtured a new gratitude for spiders after that conversation, yes, but more profound was my recognition of their lives as both connected to mine and differentiated from it. The lives of spiders had structure and integrity utterly discrete from my own.

Whenever I spotted one on my bedroom ceiling, instead of yelling for my father I silently thanked the spider, politely requested that it avoid crawling on my neck while I slept, if at all possible.

A recent study published in *Nature Ecology & Evolution* suggests that extreme weather, such as tropical cyclones and hurricanes, increases aggression in some kinds of spiders.[3] More common and more alarming is the research that shows male spiders maturing earlier in the season due to warming temperatures. Their female counterparts rely on food supply to trigger maturation. The result is a missed connection that means fewer baby spiders, with grave consequences for both the species and us.

Donika and I had, only days before noticing our spider, finished reading *Charlotte's Web*—a childhood favorite of mine—to each other before bedtime. When I tell you that I begin thinking of our porch spider as Charlotte, know that I am surprised myself at the unoriginality of this, the human instinct to not only anthropomorphize other creatures, but romanticize them with projections of our own childhood archetypes. I try not to judge too harshly the small ways we attempt to cultivate the precious among so much anthropogenic devastation, though I know these instincts to romanticize, project, and make precious have often served as justifications for that same devastation.

On the afternoon of August 10, an unpredicted storm batters the windows of our house with fallen branches. A few moments after the lights flicker off, the sky goes dark, the sun eclipsed not by another celestial body, but some nearer cloak swept in by the thunderous winds.

Not a tornado—but equally devastating—the rare derecho is known for its long duration of hurricane-force winds. For nearly an hour we sit in our dark basement as winds clocking in some places at 110 mph roar over eastern Iowa and northwest Illinois.

When the wind ceases, we stagger upstairs from the basement and fall asleep almost instantly, though it is only midafternoon, as if the storm has ravaged us in some invisible way. We awaken to the growl of chain saws, wielded by the industrious midwestern dads of our neighborhood, eager to chop the broken tree limbs.

Our spider, of course, is gone.

The economic price of damage incurred by the derecho—which is estimated in the days that follow at $7.5 billion—renders it the most costly thunderstorm disaster in U.S. history.[4] The scale of most disasters, I observe, is estimated by their cost to human economies.

After three nights without power, I insist that we check into a hotel for a night so that I can get some work done.

I jog around town and see power lines swinging in tangled snarls over the sidewalks. Enormous trees have split down their trunks and caved in the roofs of houses, crushed cars in their driveways. Weeks after the storm, thousands will remain homeless in the cities of eastern Iowa.

In the words of one climate scientist quoted in *The Des Moines Register*: "It's going to get A LOT worse."

There is so much to mourn that sometimes it's hard to discern from where my sadness springs, or to what it belongs. The new ease of our lives sometimes feels like a betrayal of those who need it more, though I know that the guilt of privilege pays no debt.

Sometimes, becoming a stranger to oneself is an opportunity

to become someone better. Over these recent months, I have become less interested in geologic time, and more interested in the people and animals where I live.

4.

During the derecho I worried about the squirrels whom I watched each morning carry mouthfuls of debris to replenish their dreys—shaggy bundles that sit in the uppermost forks of trees. What of the crane I spotted occasionally along the river as I ran? Less so the foxes, groundhogs, and chipmunks who can retreat to their dens. But most of all the small birds—warblers, tanagers, thrushes, sparrows, and finches—red and gold—who flit around our bird bath. Where does a 0.3-ounce creature go in a 100 mph wind? They knew better than me and were back at the birdbath the next day. The heavy rains didn't scare them away, but the bigger birds did.

There are three black crows who hang around our neighborhood in a little gang. When I leave for my morning run around six, they sit atop the streetlamp down the block, murmuring to each other.

Crows are among the most vulnerable to arboviruses, West Nile in particular. A 2007 study found their numbers had declined by 45 percent across the United States since the introduction of the virus.[5] They are known to recognize human faces, so I begin bringing a handful of dog kibble to scatter under the lamp as a gesture of friendship. As I shake the kibble out of my hand, I stare up at them, so that they might begin to know my face, and to trust it.

One afternoon, as I wash the dishes, the busy traffic and chatter of the birdbath pauses. In the stillness that follows, a Cooper's hawk lands on the birdbath. Her talons grip the rim as her yellow irises track the groundcover and nearby shrubs. Cooper's hawks mostly feast on small birds, which they squeeze to death and sometimes drown by holding them underwater. In the middle of the twentieth century, widespread use of the toxic pesticide DDT caused the shells of the hawks' eggs to grow so thin that they were frequently crushed by the weight of incubating mothers.

The little birds never return to the birdbath. The irony of our sadness over this is not lost on me.

Most of us think of ourselves as good people, but sometimes I trip on the fact that I am a menace to all earthly life. If I can know, as I do, that I have blood on my hands from every war my imperialist country wages, that my whiteness is implicated in every structure that protects its privilege, that most days each article of clothing I wear represents the exploitation of someone's labor, then I ought to more mindfully expand my conception of my life's impact on the world beyond the human. I'm referring less to household recycling or reusable grocery bags than to a fundamental understanding of my role on this planet. I'm thinking less of rueful jokes to express and relieve my feelings of powerlessness, and more about staying in this time frame, the one in which my existence matters.

5.

I keep waiting to miss New York, but the missing never comes. I am learning how much of my love for that city is love for my own history there. As the weeks pass, I fall in love with this place, too, and less for who I am here than for what is here with me.

I've been assembling this catalogue since our arrival, and sometimes it feels like a tiresome exercise. Some days, I would rather just indulge a romance with my new environment, be a lover of nature, be, as Emerson described, someone "whose inward and outward senses are still truly adjusted to each other; who has retained the spirit of infancy even into the era of manhood," who experiences "a wild delight . . . in spite of real sorrows."

I could talk about the prairie grass instead, though it doesn't have a face, and its sorrows are as great as those of any animal.

Don't mistake me: I do feel a wild delight at my bare feet on the ground, the chorus of owls each evening, but increasingly, the delight is shot through with grief, and that is what I am finding more precious as time passes.

Sometimes grief is not worth much to anyone but the aggrieved, but I have begun wanting to love everything the way I, until recently, loved only a few mammals: as they are, with my actions, with a stake in their suffering.

Perhaps attention is not enough, though it is something. It is the beginning of all preservation. I am looking for a way to say *I love you* that matters. Before there is nothing left to say but *I miss you*, into the wind.

Notes

1. Helge Kampen and Doreen Werner, "Out of the Bush: The Asian Bush Mosquito *Aedes japonicus japonicus* (Theobald, 1901)

(Diptera, Culicidae) Becomes Invasive," *Parasites & Vectors* 7 (February 2014): article 59.

2. "Climate Change Poses Risk to Hummingbirds, an Important Pollinator," *Conservation in a Changing Climate*, May 11, 2018, climatechange.lta.org/risks-to-hummingbirds-an-important-pollinator.
3. Alexander G. Little, David N. Fisher, Thomas W. Schoener, and Jonathan N. Pruitt, "Population Differences in Aggression Are Shaped by Tropical-Cyclone-Induced Selection," *Nature Ecology & Evolution* 3 (August 2019): 1294–97.
4. Andrea May Sahouri, "$7.5 Billion and Counting: August Derecho That Slammed Iowa Was Most Costly Thunderstorm in US History, Data Shows," *Des Moines Register*, October 17, 2020, www.desmoinesregister.com/story/news/2020/10/17/iowas-august-derecho-most-costly-thunderstorm-us-history-7-5-billion-damages/3695053001.
5. Kimberly Hall, "Climate Change in the Midwest: Impacts on Biodiversity and Ecosystems," *Great Lakes Integrated Sciences and Assessments Center*, March 2012, glisa.umich.edu/media/files/NCA/MTIT_Biodiversity.pdf.

HOW DO YOU LIVE WITH DISPLACEMENT?

EMILY RABOTEAU

The following is a diary of the first three months of 2020. During that time, the global coronavirus pandemic overtook all our attention. I abandoned my nascent climate activism to homeschool my children under quarantine, even as I understood that the two crises—climate change and COVID-19—weren't in competition. The only action I could maintain was writing down what people in my network said about what they were losing, or stood to lose, from both threats. (Some said that if the same number of people who eventually took the coronavirus seriously took climate change seriously, we might actually save ourselves.)

Climate grief and coronavirus grief feel strikingly parallel. The solutions to both problems rely on collective action and political will. In both cases, and for the same insidious reasons, the poor suffer more. In the United States, our efforts on both fronts are disabled by a reigning power that denies science and values individual liberty over the common good. In New York City, where

I live, at "the epicenter" of America's outbreak, the virus has disproportionately attacked Black and brown low-income communities already plagued by environmental health hazards. The zip codes, like mine, with the worst air pollution have also had the highest coronavirus case counts and fatalities. Many of the voices that make up this chorus come from these communities and must be foregrounded in the climate conversation that has traditionally marginalized us. It was my ambition, in gathering our voices, to suggest that the world is as interconnected as it is unjust.

1.1.2020
"Happy New Year from Stone Town, Zanzibar," said Centime, "a place of ghosts, if any exists." Rereading the canon of Black Studies, she realized that when taking field notes, the main question should be this: *How do you live with displacement?*

1.4.2020
"Whatever actions I can take to align with the earth, I must take them," said Daniel, whose drag name is JoMama, from the Bronx.

1.5.2020
"I farewelled my beautiful garden a few weeks ago. I've hung on and hung on but there just isn't the water. Farewell herb tea garden, veggies, wildflowers and carefully curated collection. All those dreams . . ." said Pen, in Queensland.

1.10.2020

A week after Trever said the sea was creeping toward the door of the house in Punaluu, Oahu, where he'd learned to fish as a boy, his family decided to sell the house, because the road of the shoreline the house sat upon collapsed into the ocean.

1.11.2020

"The weather we're getting today in NYC is a reflection of how we treat the world: Trash," said Yahdon, from Brooklyn, where it reached sixty-six degrees during the second week of January.

1.15.2020

Over chile rellenos at Posada Tepozteco, Tim complained of the air quality in Mexico City, where he lives. "The weather has become a bell jar," he put it.

1.18.2020

"It's about leaving something that will outlast us, after the people in the archive are gone, after the archivist is gone, after the world changes," said Laura, who spent four years archiving Radio Haiti after surviving the earthquake and is drawn, these days, to reading postapocalyptic literature.

"Preserving this collection assumes there will be a future, that someone will be alive to remember." What Laura remembered when she regained consciousness, trapped in the rubble with the corpse of her landlady, was the singing.

1.20.2020

Geronimo's third-grade curriculum at Dos Puentes Elementary was finally starting to confront the issue. He looked up from his reading homework on the endangerment of the monarch: "A Billion Butterflies Have Vanished."

"Why aren't people listening?" he demanded. "Will humans die, too?"

"We have to fight for butterflies and people," I told him. "That's why we gave you a warrior's name." I put the revolution handbook in the tote bag that said THERE ARE BLACK PEOPLE IN THE FUTURE, kissed him good night, and left for the meeting downtown.

1.27.2020

"You know there's something really wrong when writers join a group," joked Elliott, at the second Writers Rebel meeting. Snacking on Petit Écolier cookies, we imagined new direct-action tactics to protest extinction that didn't feel grand enough, yet were better than nothing.

Andrew, the joyous troublemaker who edited the book on creative campaigns for social change, said that common pitfalls of fledgling climate activist groups like ours include unspecific aims and ranks that are too white. Calling the problem a problem was not the same as solving it, I thought.

1.29.2020

Sunday was so windy that Officer J. politely demanded that the fifty white rebels who'd gathered to march on the pedestrian path across the George Washington Bridge break the wooden handles off their hand-painted signs, lest they fly into traffic and shatter a windshield.

"Zero emissions!" yelled a cyclist in support as the protesters walked over the bridge alongside the suicide net. They were spaced ten feet apart so that the cars entering NYC could read their signs:

FLOODS
EPIDEMICS
MEGA-STORMS
WILDFIRES
FAMINE
MASS EXTINCTION
TELL THE TRUTH

1.30.2020

Leonard said he regretted having to cancel his trip from L.A. to China because of the virus. At the same time, health and infectious-disease experts were still sounding the alarm they'd been sounding about climate change making the risk of other novel afflictions much more explosive.

Victor and I counted eight people wearing surgical masks on the subway platform at 125th Street while waiting for the A train to carry us back home after date night at Maison Harlem.

2.1.2020
Eneida said over pizza on movie night at our place that her mother, who has dementia, kept getting lost in the Bronx trying to find the airport. She wished to fly back home to Puerto Rico, though two and a half years after Hurricane Maria, her abandoned house in Cabo Rojo had been overrun by termites.

2.3.2020
Sujatha wrote in shock from Sydney about the fallout of the bushfires, the evacuation of her cousins from the outskirts of the city, and the loss of her friend's farm, where all the animals burned to death.

"The scale of the loss is incomprehensible," she said. "We've spent much of the summer indoors with our air filter on. Climate change has never felt this close."

2.4.2020
"Have you noticed your planet burning lately? Did you know that faith traditions and science both say that we need to dump fossil fuels? Come and find out more with your Congressman, Adriano Espaillat, and local faith leaders," began the invitation in my in-box. Since I had indeed noticed the planet burning lately, I sent an RSVP: Yes.

2.5.2020
"He left us indelible instructions with which to clear the way in

this here burning world," said Christian, eulogizing the Bajan poet Kamau Brathwaite, whose death we mourned today. From "Words Need Love Too":

> How to make sense
> of all this, all this pain, this drought
> scramble together vowels jewels that will help
> you understand will help you understand these
> rain.

2.13.2020
Ryan said it was T-shirt weather again in Antarctica, where the temperature on Seymour Island recently reached sixty-nine degrees.

2.15.2020
"Bureaucracies are designed to tell you no," said our congressman, the first to speak at the Faith Forum on the Green New Deal. He felt confident that once the resolution became legislation, the deal would pass the House, but not the Senate.

"The biggest problem in the district office is getting bounced around. Problems grow severe. Time does that. Poor people that contribute less to hurting the environment are suffering most."

"My eight- and ten-year-old kids are living in a world that's a degree warmer than when my mother was a child," said Allegra, born a Southern Baptist in Midland, Texas; now a NASA scientist who spoke after the congressman.

Five-hundred-year storms have become twenty-five-year storms, Allegra said, recalling the destruction of the Inwood Hill Nature Center by an almost ten-foot storm surge during Hurricane Sandy. She used to take her kids there when they were small.

"The last time Earth was as hot as today, seas were thirty feet higher." Allegra illustrated this fact with a projected slide of the 110-step stair street a mile north at 215th St., recently rebuilt by the congressman. There was a dark watermark at step fifty.

"Although sea levels are rising twice as fast in New York City as in the rest of the world," Allegra said, "hope is a discipline for survival that we may as well call 'love.' This day. This panel. This community. Our world. We're here to fight for it."

Rivka, a Talmud teacher and founder of the neighborhood Sunrise Movement chapter, spoke next, sharing the story of the firstborn child of Egypt who waged a war against their parents for refusing to free the Israelites, risking their firstborns' lives.

"We're talking about perversion of morality," said Rivka, making an analogy to the present moment. "We're being begged by our children to save their lives. Failing to meet that responsibility should revolt us. They did not ask us to be born.

"What happens when the generation in power fails to live up to the promise to protect our children? It baffles the mind. Society turns upside down and inside out," Rivka said.

Brother Anthony, a Franciscan monk, believes "our denial of the climate crisis is a fear of death. Spiritually, as a nation, we're morbid, insecure, lack control, don't trust our neighbors or other peoples, we don't believe we can change.

"We resort to tyranny and exploitation and deal death to other peoples and to the Earth so that we can hold on to our way of life and hold on to our stuff. This is futility, a useless clinging. I advocate for a re-enchantment with life and creation," the monk continued.

"Another way is possible; hence, another world is possible. To convert means literally to turn together. Time is short. I pray we still have enough time to turn together toward enchantment."

2.16.2020

"Is it in the flood plain if it's in the gray area, or the blue?" asked Victor. We'd visited a Bronx property at an open house, and he was confused about where it fell on the Sea Level Rise Map, projected for the 2050s five-hundred-year floodplain, that Allegra had shared the night before.

"If I'm understanding correctly, the house is a hair outside of the gray floodplain zone," said Victor, studying the map on the screen of his phone. "But truly just a hair."

On our way to the open house, I recalled what Rivka said. "Even though Noach was a believer, and built the ark as God instructed, he didn't really comprehend the catastrophe until he saw the floodwater with his own eyes. Like Noach, we must act as believers, even as we don't really yet comprehend."

2.23.2020

"You can't find wood this strong anymore," said Derek, the home inspector, caressing the beams in the basement of the house that may sit in a future floodplain. "They chopped down all the trees with this many rings a long time ago. They don't exist anymore."

3.1.2020

At the climate storytelling workshop, Adam, an atmospheric scientist, admitted bedevilment by language when asked by media to attribute particular hurricanes to climate change. Like Spock, he cannot lie, and as a result, sounds craven. *Wimpy* was the word he admitted to.

"Why can't scientists say the simplest thing?" Emily, a journalist, demanded to know why Adam could not also speak about extreme weather as a citizen, as a parent, as a person admitting to fear. "Why aren't you more stark?"

"I don't have an answer," Adam deflected. He worried that words like *crisis* and *emergency*, when paired with talk of climate, would lead to authoritarian government action.

3.1.2020

"Bring your own plastic bag!!! Traiga su propio bolso de plastic," announced the sign on a bodega at 181st Street.

3.2.2020

Sophie said she spent three months crocheting a "show your stripes" blanket too large to fit in the photo of it, which featured her napping cat for scale. "Its colors show global average land ocean temperature changes between 1880 and 2019," she said. Crafting it had calmed her terror.

3.5.2020

"Oh, you can't trust those flood maps," Meera warned over the squash soup I'd made for lunch after I mentioned the house we'd fallen in love with. "If the city was truly honest about the properties at risk, we'd have another mortgage crisis on our hands."

3.6.2020

"Headquarters hasn't told us anything yet about the bag ban," said Jason, the sales associate at Party City, where I bought trinkets to fill the goodie bags for our son's seventh birthday party. He corralled a dozen helium-filled balloons into a giant plastic bag.

3.9.2020

"Just finished my apocalypse shopping," said Honoree, whose stockpile included dried fruit and nuts, ground coffee, plant milk, popcorn, olive oil, vegan sausages, unsweetened peanut butter, seventeen gallons of water, and two dozen eggs. "Better safe than sorry."

3.11.2020

"Maybe God is tryna tell us something?" suggested TaRessa, in Atlanta, after satellite space images showed a drastic drop in Chinese air pollution after the novel coronavirus shut down factories there weeks ago.

3.12.2020

"I feel scared," said the woman waiting across from us at gate D10 in LaGuardia airport, slightly embarrassed when she caught me watching her wiping down the armrests and the seat of her chair with Lysol. The airport was nearly empty.

3.13.2020

"We are all scared," Harouna said, at the empty Pain Quotidien on 44th Street. Other servers were hastily removing tables and chairs, having just been ordered by corporate to thin the density and decrease the risk.

Over drinks at the Garrison on Wednesday night, Ayana, who had a painfully sore throat, said she was scared about how to get to her aged mother in Philadelphia in the event of a lockdown, as well as for her mother's health.

Dr. Stephen, an Upper West Side dentist, predicted that when they finally closed the public schools in NYC, the looting and chaos would begin. I wanted to argue, but he was drilling into my tooth.

The dentist proudly showed me his stockpile of surgical gloves, face masks, and hand sanitizer. He would only treat patients now after taking their temperature, he said.

The day before the World Health Organization declared the outbreak a pandemic, Paisley, the poet laureate of Utah, said she seriously did not understand anything anymore. "Every day reading the news, it's like my whole life has prepared me for nothing."

"My middle of the night revelation was that it is a privilege in this country to say you actually have Coronavirus cuz that means u have access to a test that confirmed it. U have the means to get one, wealthy, and or connected," wrote my neighbor Sheila, a diehard Prince fan.

"All the rest of us who have any of the whole range of symptoms that are basic to a cold or allergies or flu cannot really know what it is cuz we don't have the connections of this devil ass administration and their rich friends," Sheila fumed.

"Truly, fuck this no-universal-healthcare-having country right now," she said, right before Trump declared a national emergency—one in which his administration claimed not everyone needed to be tested. "Ese diablo cabrón mentiroso jámas ha sido mi presidente."

3.13.2020

"Is Grandma going to die from coronavirus?" asked Ben, our seven-year-old, who was helping me mix the filling for a sweet potato pie while Victor tried to reach his mother in the ER.

Miranda said she was pulling her kids out of school because she felt it was the ethical thing to do.

3.14.2020

"No one knows what'll happen over the coming weeks in Lagos, Nairobi, Karachi, or Kolkata. What's certain is that rich countries and rich classes will focus on saving themselves to the exclusion of international solidarity and medical aid," said Centime, coughing over mint tea with honey. Her breast cancer was back, had spread to her spine and lungs.

While ER doctors in northern Italy were warning us in the States to change our behavior to flatten the curve of contagion, Paolo said the social isolation in Turin was making his aged parents lose their minds.

In Bennett Park, Sula's mom, an NYC public school teacher who sat a social distance on the opposite side of the bench from me while our children played soccer, said that she and her colleagues were desperate for the schools to close.

3.15.2020

In Fort Tryon Park this afternoon, Jake's mom angered him by snatching away his walkie-talkies before he could share them with his friend, because she feared spreading germs.

The other Ben reported from lockdown in Barcelona that "for the last two nights the entire city has gone out onto their apartment balconies at the same time to applaud for healthcare workers and declare their own vitality and solidarity and stubborn joy."

Leaving the building to stockpile groceries and goods with his anxious wife and toddler, my neighbor Emmanuel, an NYC tour guide, said he was laid off on Friday and planned to apply for unemployment benefits.

"I knew this was coming but I thought we had more time," I said to Victor when the mayor announced school closures.

"Anyone who would like a grab-and-go breakfast may pick it up at 7:45 and lunch at 11:00," said Principal Tori of Dos Puentes Elementary School, via robocall. The schools were provisionally scheduled to reopen April 20. "There is food for anyone who needs it," she stressed, her voice breaking.

"When is the coronavirus going to go away?" asked Ben, at bedtime. "Is it possible that it will never go away?"

3.16.2020

"What do you do with a national emergency that requires community action, in a country run by white people who not only do not believe in community, but have spent all of history trying to destroy yours?" pondered Hafizah, en route back to Brooklyn from L.A.

"Watching the spread of this virus has sort of been like getting fitted for new glasses—you start to see a bunch of things more clearly and catch sight of previously overlooked beauty while also recognizing the ugliness in some of what formerly impressed you," observed Garnette, in Charlotte County, Virginia, who was having difficulty taking deep breaths.

Ayana's octogenarian mother said she had seen some things in her lifetime and knew that in times of chaos, like this, we mourn what we're losing, dwell on loss, and lack the imagination to see that what we're losing may be replaced by something better.

Nelly drove home down the Taconic between "a waning gibbous moon to my right, an eggy sunrise to my left, with Langston Hughes's *I Wonder as I Wander*, him wandering through Haiti, Cuba, the South, Russia, me wondering how these times will change time, how this time we might change."

3.17.2020
From Amsterdam, Nina reported that people in the Netherlands all applauded the first responders and essential service providers tonight at 8:00 p.m.

"Shelter in place?!" cried Alicia, an incredulous elder. "How will they enforce it? This is New York. Folks won't listen. We're not in a prison or a concentration camp. I'll be damned if anybody tells me I can have a visitor or not. Unless you're paying my mortgage, don't you dare." Katherine visited her parents in New Orleans, where the Black poor were stranded in Katrina, and stood six feet away in the driveway. "My son tried to hug my mom. When we reminded him he couldn't, she went inside and came back out with one of those long, grabber tool things at the end of which was a piece of paper on which she'd written 'hug, hug, hug.'"

3.18.2020
Briallen, who previously had a large chunk of intestines removed, posted a picture of the votive candles in jars arranged in her windowsill. "I shat myself while praying this morning," she said, before heading to a pharmacy in Elmhurst, Queens, to stock up on Depends.

Looking at the red spread of the virus across the world map, Damali, from Uganda, said she was relieved for once that Africa wasn't the seat of calamity.

3.19.2020
Adam, who works in fundraising for NY Presbyterian Hospital,

said a rich banker donor contacted him to ask for the "special" number to call for testing, and that the hospital was looking at the former quarantine islands around Manhattan as pandemic real estate.

"God has given us to one another so we can care for one another," wrote Pastor Jon—with whom I last spoke in February at his office in Cornerstone Church, where, over weak coffee, he prayed for me in my "bewilderment" about the climate crisis—and who is now quarantined with the virus.

3.20.2020

"Fear and panic do not lend themselves to an empowering homebirth," cautioned Kimm, our former midwife, regarding increased interest in homebirth among pregnant women anxious about going into the hospital during the pandemic.

Emmanuel's wife, Rebecca, a juvenile defense attorney, said she argued her client's case over the phone instead of going to court.

Rivka said, "We canceled seder, which is difficult to overstate what a big deal that is, culturally. It's our son's first Passover ever, so it's sad. My wife and I will be doing seder alone in quarantine."

3.21.2020

Aysegul, suffering a fever and shortness of breath while quarantined in Paris, said, "All around the city I know what my friends are having for dinner this evening: roast chicken, coconut–butternut squash soup, potato salad, tofu and ginger noodles."

Lamenting the end of the age of dinner parties, Angie said it was also only a matter of time before the drones keep us on lockdown in NYC as they are doing in France.

3.22.2020

Today in Washington Heights's J. Hood Wright Park, Adam said laboring pregnant people could no longer have their partners with them during birth at the hospital he works for, whose towers were visible beyond the pink weeping cherry tree.

"Today we were supposed to have a memorial service for my father, which we canceled a couple of weeks ago when we saw what was coming . . . I don't know that I'll have the space to take a deep breath and just miss him. But I do miss him," said Sonya.

3.23.2020

Tony evacuated the city on Friday amidst attacks against fellow Asians for causing the so-called Chinese virus.

"I know many of us would love to be lied to. To be reassured 'this will be over soon.' That 'things won't be different.' But the truth is, things have already changed—and will never be the same," said Yahdon.

"If anyone can spare N95 masks for our birth team so we can use them for attending births and keeping ourselves virus-free, we and the families we care for would really appreciate it," said Kimm.

3.25.2020

"Preparing for the days to come to keep the family safe," said Imani, in Brooklyn, who'd sewn fashionable handmade Ghanaian wax-print face masks.

"Eventually, does the whole world go away?" asked Ben, while watching *Totoro* after a morning of home school. "Like if everyone dies?"

"Yes, viruses spread very quickly in NYC, clearly faster than anywhere else in the U.S. But so do arts, trends, ideas, passions, cuisines, and cultures. If you could map those things, they would look exactly like these pandemic maps, but they would be pandemonium maps," said Adam. "And that is why I love living here, and why I consider myself so lucky to make a life with my family here. I love New York."

3.26.2020

"Seems like almost overnight the alienating quiet of these Brooklyn streets has been replaced with basically nonstop sirens," said Mik.

"See you on the other side, brother," said Victor to our neighbor, the other, other Ben, who gifted us with a carton of eggs and asked us to take in his mail, before fleeing the city with his wife and two kids. "Stay strong."

3.29.2020

"What's going on?" asked an alarmed stranger from his parked car across the street on Friday at 7:00 p.m. We were clapping, hooting, and hollering with full lungs from our front stoop, along with the rest of NYC, for the essential workers on the front.

On Saturday, Amy from Apt. 5, single mother of twins, gave me her keys and said we were welcome to eat what was left in the fridge. Before nightfall, she fled the city for Connecticut in a rental car, afraid of the president's threat to close Tristate borders.

John said, "The weirdest symptom has to be how it wipes out your sense of smell. Drinking coffee now. It's black, but I can't get a whiff of its scent. I've had it for almost ten days. Can't even smell the rain."

4.2.2020

"You could run from Katrina. You can't run from this," said Maurice, in New Orleans, where Mardi Gras may have served as an amplifying petri dish for the virus. Going-home ceremonies with second lines to send off the dead are now banned.

4.3.2020

"So many folks are looking forward to summer as a break from COVID-19," said Genevieve, who felt economic recovery must be centered on decarbonization. "But I'm terrified the virus won't have abated and we'll also be faced with the heat waves, fires, and storms that now fill the months July to October."

4.4.2020

"Yeah, this is how it's usually done with Black bodies considered expendable. Black lives worthless except in service to commerce or science," TaRessa said. On live TV, two top French doctors had recommended testing coronavirus vaccines on poor Africans.

Centime, who was feeling weak with a collapsed lung, said, "We need to stop fucking around with theory and say that capitalism, with its industrial body and crown of finance, is sovereign; carbon emissions are the sovereign breathing; 'make work' and 'let buy' must be annihilated; there is no survival while the sovereign lives."

FASTER THAN WE THOUGHT

OMAR EL AKKAD

I grew up in Qatar, a tiny peninsula off the eastern coast of Saudi Arabia. Less than a century ago, before the boom, it was a desolate corner of the world, home to Bedouin tribes, shepherds, fishermen, and pearl divers. Today it is, by virtue of its massive oil and gas deposits, the richest country on Earth.

As in Alberta, Texas, and countless other little empires of extraction, money has lubricated a wild and rapid urban transformation in Qatar. The vacant lot where I snuck my first kiss is now a maze of sail-shaped skyscrapers. A playground where I learned to ride a bicycle has been cleared to make room for five-star hotels. The whole area—playground then and hotels now—stands on land originally reclaimed from the sea, in a neighborhood whose old Arabic name, roughly translated, means "the burial site."

If the speed with which this transformation happened is unique, the trajectory is not. In the age of capitalism everything is a placeholder for its more lucrative replacement. And is there more universal an expression of nostalgia than to return to the site of your first kiss and find it unrecognizable? Time moves this way.

But there is now another kind of obliteration, another kind of burial. Within the next century, possibly within my lifetime, Qatar's landscape will become uninhabitable. There exists a scientific definition, based on heat and humidity, of what constitutes a survivable climate. If carbon emissions continue at their current pace, it will eventually become impossible to live in much of the Middle East without constant air-conditioning. It will become impossible to exist, for more than a few minutes, outside. Sometime within the next century, stories of life in this place—the stories that constitute almost the entirety of my childhood—will sound, to new generations, like fiction. The tether between what is and what used to be, constantly stretching under the weight of history and progress, will not stretch any more. It will snap.

The axis along which almost all climate-change anxiety orients is, by necessity, pointed toward the future. It is a space that will never arrive, and because of this we are all prone to afford it endless possibility. Never mind that even if we were to impose a total prohibition on fossil fuels tomorrow the atmosphere would continue to warm for another century at least, never mind the glaciers already disappeared, the coral already dead, never mind all the damage we've done—the future is and always will be thought salvageable. We have framed climate change as a crisis of the future because its worst ramifications are still to come, and because the future is something we feel we can still control.

But we should also spend more time thinking about how climate change will upend our past.

•

This summer, by the banks of the Tigris River in Kurdistan, archaeologists found a palace, long ago drowned. It was built during the reign of the Mitanni Empire, a little-known civilization that once ruled much of northern Mesopotamia, about 1,500 years before Christ.

From a researcher's perspective, the finding is a good thing, a key addition to the knowledge base, but how it came about is the result of something far less positive. A depletion.

About forty years ago, this part of northern Iraq was flooded after the construction of the nearby Mosul dam. But so severe is the current drought that the Tigris recently dropped away to reveal a new topography, and in that barren wash the archeologists found a structure hidden for millennia—a figment of memory violently resurfacing, the way old scars reappear on the skin of malnourished bodies.

We are about to lose so much. The next few decades will see the disappearance of countless glaciers, coral reefs, animal and plant species, miles and miles of coastland. The places in which our stories took shape will become entirely changed, and the changes will birth new stories, a new telling of forgotten pasts—ghost towns rising out from the bottom of vanishing lakes, secret histories written into the rings of severed trees. In this moment of monumental import, this apex of a turn beyond which lies our survival or eradication, climate change is going to render our past as unrecognizable as our future.

As I write this, the Brazilian government is passively aiding in the wholesale destruction of the world's largest rainforest—the source of a fifth of this planet's oxygen—in the name of commerce. The president of the United States and his corporate and political enablers are busy tearing up environmental regulations, knowing full well they will never have to answer to the future generations of children who'll struggle to breathe every day as a result.

We can imagine—we should imagine—the repercussions of this wanton cruelty and greed, the ways in which they will reinforce all our existing inequities. The rich will move to the temperate, less-ruined places; the poor will be made refugees and the influx of refugees will prompt bloodshed at the borders of the privileged world. We will continue to kill one another over food and over land as both become scarcer.

But we will kill one another over memory, too. There's a reason terrorist groups such as the Islamic State refuse to use the present-day names of the places in which they operate, adhering instead to anachronistic markers of long-gone empires. It is, in its own way, a kind of diseased relationship with memory, this notion that the only acceptable ordering of the Earth is some previous ordering of the Earth, some other set of lines drawn on a map. But it is nonetheless a powerful notion, and in the coming decades we will render millions and millions of people more susceptible to it. In the great undoing that is climate change, there is born an entirely new means by which one dispossessed generation can say to the next: your home used to be here and it was taken from you unjustly. It was the sea or the storm or the drought that buried it but it was something else that killed it—its killer was a system, an institutionalized greed, an infinitesimally small bump upward in the wealth of the already wealthy.

So many extremists will be made this way, and when those extremists inevitably unleash whatever evil they mistake for revenge, not one of us will be able to honestly say we didn't see it coming.

It may be the case that we're past the point of fixing this. Perhaps we might still be able to mitigate the worst effects of climate change, but maintaining our current ease of life into the next century and beyond is optimistic to the point of hallucination. To accept this outcome is difficult, because it entails accepting that the future is no longer a space of infinite possibility—rather a house mortgaged to the hilt, a foreclosure in waiting.

We must create new ways to think about what comes next, but also about what came before. As coastland drowns, as wildfires thin the ground and thicken the air, as changes that used to take centuries begin to take years, it will become increasingly difficult to anchor our memories to a geography, to a stable piece of land. So we must find other anchors—anchors that link memory to people, to relationships, to the solidarity and compassion and resistance that will serve as our only useful lifeboats in this storm.

Some of this work will fall, as it must, onto the shoulders of historians, researchers, people who make it their life's mission to measure the depth of marshland and the health of honeybees. But more so, it should be the work of artists, of storytellers. Memory is a fiction after the fact, less an accounting of the moment than of our passage through the moment. We have an obligation to document and preserve this compendium of fiction, these stories we tell ourselves. And we have an obligation to do it now, meticulously, because the stories and the empathy they engender might save us still, might move people to act. And because in time the world in which those stories took place may well vanish, in its place a different, emptier world—emptier of nature, emptier of life; the stories, once lost, lost forever.

There will come a time when millions of people around the world—from the islands of Fiji to the coasts of Bangladesh to the southern tip of Florida—can no longer point to a brown spot on a map and say: I am from here. There will come a time when many sense triggers of memory—the smell of petrichor, the sound of a particular birdsong, the feel of snowflakes against the skin—will be more and more fleeting, and in turn the things they once triggered will burrow deeper and deeper into the back of the mind. It is already happening.

We must learn to become conservationists of memory. Otherwise, this damage we have done to our planet will cost us our past, as it may have already cost us our future. And without a past or a future, what are we? Nothing. A flickering violence of a species, here such a short time, insatiable, then gone.

UNEARTHING

LIDIA YUKNAVITCH

Bombs Bursting in Air

A lot of people have heard of the Manhattan Project, the research and development activity during World War II that resulted in the first nuclear weapons, spearheaded by the United States. The project is lodged within American culture by name partly because unimaginable destruction also engenders endless fascination. Many people have also heard the name Robert Oppenheimer, the director of the Los Alamos lab where the actual bomb was designed. There is an entire mystique that has grown around Oppenheimer. Around the Manhattan Project. From the fictional Doctor Manhattan in DC Comics back through early films and novels that took the program as their source material, the story of how the United States came to create and use nuclear bombs has buried itself in our collective unconscious. The Manhattan Project is at the heart of the dark satirical movie *Dr. Strangelove.* The Oppenheimer phrase "Now I am become Death, the destroyer of worlds," language that has been sourced to the Bhagavad Gita, has also transcended its origins.

Fewer people know the story of the Hanford Site in Washington State, where, beginning around 1943, uranium was irradiated toward the production of plutonium, a key substance in Oppenheimer's work. Hanford was part of the Manhattan Project. The site was also known by other names between World War II and the present, names that hinted at its nuclear ties: Hanford Project, Hanford Engineer Works, and the Hanford Nuclear Reservation.[1]

From 1944 to the present, Hanford officials have continued to tell a story that their operation does not pose a health threat to the surrounding land, or animals, or workers, or to people anywhere. The story changes over time, the shape and flux of the explanations, the rate of cleanup, the promise of containment or removal. But the site is still radioactive, the land and water and animals still carry contamination, and the workers, their families, and surrounding civilian communities are still getting sick and dying differently than they might have otherwise. Our bodies carry everything that has ever happened to us, the way the land carries everything of humanity.

I've lived in Washington and Oregon for most of my life and still know many locals who do not know the city's history. If I say "Hanford" out in the world beyond Washington and Oregon, there is even less recognition. But history continues to reassert itself here. At present, the site is overseen by the Department of Energy and a contractor, Washington River Protection Solutions. The current cleanup of more than 60 million gallons of chemical and nuclear waste stored in underground tanks is expected to last well beyond fifty years, maybe one hundred, in addition to the years the waste has already lived.[2]

During World War II, the Hanford Site produced the materials needed to build the country's nuclear arsenal. The land near the Columbia River was requisitioned away from the people living there by the U.S. government and military. Farmers and other civilians received letters telling them they had one month to leave. Only a few were mildly compensated. Native Americans, specifically the Confederated Tribes of the Umatilla Indian Reservation, the Confederated Tribes and Bands of the Yakama Nation, and the Nez Perce Tribe, who all lived in the Columbia Basin, were not compensated at all. They were, however, given "visitation rights"—they could arrange to go up to White Bluffs Island by truck to fish. The Wanapum tribe, who used to live on the land year-round, were forced to relocate to Priest Rapids.[3]

The workers in charge of the current Hanford Site cleanup and maintenance are still getting sick, just like the workers and residents of the last seventy-seven years. The dead who were poisoned, those who contracted diseases through contaminated water, those who developed deformities and fatal chronic diseases—no one outside of their communities knows their names.

Not like the name "The Manhattan Project."

The first nuclear devices detonated at Hiroshima and Nagasaki were named Fat Man and Little Boy.[4]

A Boy Story

I knew a boy who lived near Hanford, Washington. In 1976 we were both competitive swimmers. He was a year older than me. I was thirteen. I lived in Bellevue, Washington, quite far from Hanford, but I'd see him at swim meets in Washington and Oregon. Most often, I'd see him in the summer at the Wenatchee Invitational, a swimming event about two hours from Hanford. Wenatchee hails as the "Apple Capital," and the fruit was emblazoned on the front of the swimming medals, the awards as big as our hands. Faux gold, silver, and bronze with big chains. All the kids loved wearing them. They were my favorite medals. I still have a few, though they are smaller in my hand.

The boy and I were both very fast in the same event: the hundred-yard breaststroke. Maybe that's why we were friends of a sort. Like friends at a distance. Or friends whose bodies carried something similar. Sometimes we would have a small conversation about races or times or apples. Sometimes we'd just exchange smiles or lock eyes right before one of us stepped toward the starting block. His family lived somewhere between Kennewick and Hanford, downwind from Hanford, where his father worked. In the Pacific Northwest, the Downwinders are people who have been exposed to radioactive contamination and ionizing radiation that has entered groundwater, food chains, air, and drinking water.

He had swimmer's hair—you know it? A sheen that overtook the brown.

Green eyes. Extra-broad shoulders. The elongated torso of a swimmer.

His name is in my body, like I've swallowed an apple whole.

Art

When I was thirteen my father took me to see a showing of *Dr. Strangelove* at a theater in Seattle, Washington. My mother may have been there too, but I don't remember her with us. Ghost woman gone to drink. These were dark years for me with my father. He was omnipresent in the worst ways, magnified with rage and violence, and I'd hit puberty, an event horizon if you are the daughter of a man who abused the women in the family. But somewhere inside him he carried the trace of an artist. He was an architect; he loved literature and film and classical music and stand-up comedy and great television. He loved Jacques Cousteau and *National Geographic*. He hated Republicans. Terrible things and beautiful things can live inside the same person.

I loved *Dr. Strangelove* immediately, but I didn't yet have the vocabulary to explain why my imagination latched on to it. Something about violence. Something about art. Something about satire even though I don't think the word *satire* was in my lexicon yet. I laughed but didn't know if it was okay to laugh.

At home after the movie, we ate dinner and my father talked about the Cold War. I knew a little about what the Cold War was from school. My father talked. I was thinking about nuclear weapons.

When my father was about twenty, he said, he and his friend had jobs at some plant where spent fuel rods from a nuclear reactor were stored in giant water tanks. At that time, he'd recently quit school at John Carroll University, a Catholic school in Ohio. His friend Eddie Garilitus had swum in one of the tanks. In my memory of the story, my father recounted swimming in one, too. But I recently asked my sister, who is eight years older than me, and she thought no—just this Eddie guy. I remember thinking, holy shit. Chernobyl had not happened yet, but would in 1986.

Was it a dare between men?

Something to pass the time at a boring job?

Did radiation get inside the body of the male swimmer?

What might have been born in my father that turned his rage nuclear?

Mind over Matter

For the duration of the Cold War, nuclear technology and plutonium processing evolved at breakneck speed.

Safety procedures and waste disposal practices did not.

For over seventy years, Hanford's reactors released radioactive materials into the air. Into the Columbia River. Into the aquifer and ground. Into the bodies of everyone who worked there or lived anywhere near it, as well as the area's animals and plants.

In 1986, formerly classified documents became available through the Freedom of Information Act. There was one experiment called the "Green Run," documented in a report called *Dissolving of Twenty Day Metal at Hanford*, in which officials at Hanford released 7,780 curies of iodine-131, as well as 20,000 curies of xenon-133, into the surrounding atmosphere within seven hours. I know. Those numbers are difficult for civilians to understand. Just for comparison, the Three Mile Island accident released between 15 and 24 curies.[5]

The language used to describe the cleanup effort is almost lyric in its epic failure: Weapons production reactors were "decommissioned" after the Cold War but left behind around

53 million U.S. gallons of high-level radioactive "waste." The leftover material was stored in 177 tanks, including some 25 million cubic feet of solid radioactive waste. Underground tank "farms" emerged. Over the years, some of the tanks began to leak through their "single-shell walls" and into the groundwater and river and air. In 2011 the DOE pumped much of the liquid waste into twenty-eight newer "double-shell tanks." Those leaked too.

The planned construction of the waste treatment plant has been repeatedly delayed for decades now, and the cost of construction and cleanup keeps rising.[6] Different administrations have conflicting ideas about what should happen next. The ground and the water and the buildings remain alive with radioactivity, the water threatening to migrate into the Columbia River, metastasizing, stretching out like the fingers of a hand through seep and plumbing.

The holding tanks don't hold. Neither does the language describing the cleanup. The people and animals there die.

In a 2016 essay for NBC News, Ronan Farrow and Rich McHugh reported on the Hanford Site and the safety concerns of the workers. They interviewed a local neuropsychologist, who reported on having evaluated twenty-nine people with "both respiratory and cognitive symptoms," including some of the "worst cases of dementia in young people" he'd ever seen.

Farrow and McHugh interviewed twenty workers.

One woman experienced shaking all over the right side of her body.

Another man had nerve damage that sometimes made him pass out.

One man, who is now dead, was losing his memory and having trouble breathing.

All of the workers were being told the workplace was safe.

Former workers from two decades past said they were not allowed to wear protective work gear, like air tanks and other safety equipment. The workers asked their immediate superiors. The workers told their unions. The unions begged the owners and contacted shareholders. Over and over again they were told that the workplace was safe.

Hanford is the most contaminated nuclear site in the United States.

In 2015, it was also designated part of the Manhattan Project National Historical Park alongside Oak Ridge and Los Alamos.[7]

Eating Apples

I used to faint in the sun. White hair, blue eyes, fair skin. My swim coach would sit me in the shade at outdoor competitions with white towels over my head. I passed out often, especially at the meets in Wenatchee, where summer temperatures could get as high as 114 degrees. Other swimmers teased me, made fun of me for it. But swimmers are a pretty tight squad, so they'd also bring me cups of water.

My dad gave me big round salt tablets about the size of my eye. I think they were supposed to help ward off dehydration? Like Gatorade? They kind of creeped me out, mostly because they came from him.

One time this friend—the boy I knew from Kennewick—sat underneath the white towels with me and we shared a foam cup of crushed ice. Our races were always near each other. Because we were kind of sitting shoulder to shoulder, I could see a weird-looking bump on his shoulder. About the size of a golf ball.

I stared pretty hard at that bump on the boy's shoulder. Underneath the towels. Shoulders are a big deal for swimmers.

If you've ever swum breaststroke competitively, or seen someone do it on TV, you've seen their shoulders emerge from the water with great force and skill. You've seen them pull their bodies forward by bringing their arms close to their chests underwater, and you've seen their upper bodies rise like a graceful new species of water creature, then dive back down following the point of their hands and arms, a beautiful wave making a hydrodynamic curve over their nearly submerged heads.

This boy won nearly every race I ever watched him swim.

So other people were watching the bump on this boy's shoulder too.

In 1976, workers were not being told much about safety hazards at Hanford.

Neither were Native Americans, whose land still called their names.

Or farmers.

Or swimmers.

Or mothers and fathers.

Or families.

Or boys.

Animal Farm

In the 1950s, Hanford expanded its animal testing and research experiments. First, they used fish. The largest program used sheep. In a project that lasted more than a decade, scientists put concentrated radioactive iodine into sheep feed. Among the sheep receiving high doses of radioactive iodine, some developed malignant tumors. None of the ewes bore any lambs.[8]

Pygmy goats and cows, meanwhile, were used to determine how radioactive iodine traveled through mammalian milk. The scientific research had not caught up yet to the idea that radioactive contamination does not obey any boundary. Radioactive material leached mercilessly into the ground, the groundwater, plants, animals. Children born downwind of Hanford consumed the milk of exposed cows for decades.

Dogs were also used to test the effects of inhaling radioactive particles. Hairless pigs were used to try to figure out what would happen to the skin of human soldiers on a nuclear battlefield. Miniature pigs developed leukemia and bone cancer.

I've read about a man-made irradiated pond where maybe fifty-five alligators were kept. I can picture them, those prehistoric and toothed creatures seething under heat lamps. One article claims that the alligators "would hold their breath when they first smelled ether that was used to put them to sleep so researchers could conduct tests. Alligators can hold their breath for up to two hours."[9]

The alligators escaped once in the 1960s. Most of them were supposedly recovered.

To further study the effects of inhaling radioactive materials, two baboons were flown in once, but they escaped their shipping crates at Los Angeles airport, and when they arrived in Hanford, "One jumped in a water fountain and splashed the people who walked by, and the other climbed to the rafters and raised holy hell for several hours."[10]

In 1970 the radiological animal testing program was scaled down. In 2007 Hanford workers found and removed forty thousand tons of animal carcasses, shit, and other waste from giant burial trenches at the former experimental animal farm. A railroad tanker car had been stuffed with carcasses and buried. The tanks were filled with cats, dogs, cows, sheep, pigs, goats. Apparently when they tried to incinerate them, it didn't work; when workers looked inside the tanker car, they found no sign of ashes, just animal carcasses wrapped in plastic, rotting.[11]

Radionuclides were found in ducks, geese, deer, and jackrabbits at Hanford, but they were also found in animals as far away as California and Alaska.

The Columbia River is North America's largest river that flows into the Pacific Ocean. Before it became the most radioactive river in the United States, it ran like a vein through the body of the land, giving life to animals, peoples, farms. The native salmon runs . . . well, they used to be something. The Hanford Site was chosen partly because of the Columbia's abundant water, so it could cool the temperatures created during reactor operations to facilitate plutonium production.

Radioactive contamination showed up in fish, waterfowl, algae, insects. All the way into oyster beds in Oregon.

Native American communities were at the highest risk of contracting cancer, since their diet consisted of Columbia River contaminated salmon. As Robert Alvarez pointed out in his article

"The Legacy of Hanford," according to a 2002 study conducted by Indian tribes and the Environmental Protection Agency, "tribal children eating fish from the Hanford Reach have 100 times the risk of immune diseases and central nervous system disorders as non-Indian children," and the risk of "contracting cancer among tribal people was estimated at 1 in 50."[12]

Humans swam in this water, children played on the shorelines. In the 1960s, families preferred swimming near Hanford because the water was warmer, an effect of radioactive contamination.

Atomic Man

In 1976 a blast accident happened at the Hanford Site that irradiated a technician with five hundred times the occupation standard. His name was Harold McCluskey. I know that because I eventually learned about it from the news and at school. People called him the "Atomic Man."

When I say "Harold McCluskey" to people I know from the Pacific Northwest, they get a look on their faces like maybe they might know who I am talking about, but then the look mostly fades. Many more peoples' faces light up when I say "Doctor Manhattan."

When the accident happened, the room Harold was in was a space used for recovering a radioactive by-product of plutonium. After the blast, Harold's body was covered with blood. He was dragged from the room, put into an ambulance, and sent to a decontamination center. He was moved to a steel-and-concrete isolation tank by remote control, his body too contaminated for human touch.

Doctors had to remove tiny pieces of metal and glass from his skin.

Nurses bathed him and shaved him—every part of his body—every single day for months.

Harold lived and was released back into his community, but he suffered severe health problems the rest of his life. Friends and community members often avoided him.

In 2008, the Department of Energy and the contractor CH2M Hill Plateau Remediation Company began preparing the Hanford Plutonium Finishing Plant and the "McCluskey Room" for demolition, which has since been completed.

There's another reason I know about Hanford, and why that knowledge never left my body. My father died years ago, but his company has been in the news my entire life. CH2M Hill won the contract for cleanup at Hanford. In 2019 there were several incidents in which workers' skin was contaminated with radioactive waste, and the DOE said that the company needed to improve its radioactive and chemical exposure assessments, but one year later, in 2020, CH2M Hill Plateau Remediation Company was awarded $7.8 million by the Department of Energy.[13] In 2019, the DOE released a report that stated the expected cost to complete the Hanford cleanup had tripled—again—in three years. Somewhere between $323 billion and $677 billion.[14]

My father was an architect who worked for an engineering firm called CH2M Hill. It's funny how life carries echo effects and the traces of all the things that have happened all around you.

My father told me the story of the "Atomic Man" at dinner one night. So many years later, it's become as buried in me as his rage and violence.

Dead Family

My swimmer friend showed up at the "Apple Capital" Wenatchee Invitational one more year. That next summer.

I didn't faint that year.

I didn't have to sit underneath white towels.

He didn't bring me a cup of crushed ice.

He didn't win his race, either.

The golf-ball-size bump on his shoulder had grown, like a third shoulder, and everyone sort of pretended not to notice . . . but of course everyone did.

I've had skin cancer twice in my life. One time it was on the bridge of my nose—a little red scab or sore that just wouldn't heal. For the longest time I thought it was just that place where my glasses were rubbing. Everyone around me could see it. Students, fellow teachers, my husband. Sometimes it bled. Until my gynecologist sort of insisted, no one anywhere had ever said to me, you know, maybe you should get that checked out?

Another time I had a basal cell carcinoma removed from my back.

Who can see their own back?

A fellow swimmer probably could have.

I didn't live anywhere near Hanford. Hanford is about three hours from Bellevue. Upwind a long way.

I don't know why I never asked the boy from Kennewick about the bump.

I haven't been back to Wenatchee since the last year I saw him.

I still swim, but at fifty-six years old, my shoulders ache. I wonder sometimes what they ache from. Or for.

Art

In 1983 I saw the movie *Silkwood.* I was twenty years old, in my second year of college, before I flunked out. In it Meryl Streep plays Karen Silkwood, a nuclear whistleblower and labor union activist who was maybe, to my mind probably, murdered—she died in a car crash while she was investigating dangerous labor practices at the Kerr-McGee plutonium plant where she worked. I cried so hard at the end of the movie I couldn't breathe.

I had a kind of echo image in my body already—the Hanford story, the Atomic Man, my father's rage . . . the bump on a boy's shoulder.

Karen Silkwood had forty times the legal limit of radioactive contamination in her body.

My son is nineteen. He's never heard of Hanford, though he's heard of the Manhattan Project. He's never heard of the movie *Silkwood* either, but that's about to change. His father and I are progressive parents who are also educators. It seems time to show it to him, even as time right now is carrying a tumor, an echo effect of other times, different fears, different brutalities. His favorite movie of all time, so far, is *Dr. Strangelove.*

Shoulder

The swimmer boy from Kennewick died.

Two of his brothers got thyroid and liver cancer, respectively, and they died, too. His sister developed a brain tumor and died. His father didn't live to see fifty-five years. I don't know about his mother. I hope she left. Kennewick is forty-one miles from

Hanford by Route 4. If she stayed, it's in her body too. Everything of Hanford. And a spent grief that no body could hold.

I just carry a story in my body, and a will toward making stories. I believe my will to make stories comes from my unwillingness to let things lie.

Something about land and animals and the bodies we bury.

Something about unearthing.

Notes

1. "Science Watch: Growing Nuclear Arsenal," *The New York Times,* April 28, 1987, www.nytimes.com/1987/04/28/science/science-watch-growing-nuclear-arsenal.html.
2. Ronan Farrow and Rich McHugh, "Welcome to 'the Most Toxic Place in America,'" NBC News, November 29, 2016, www.nbcnews.com/news/us-news/welcome-most-toxic-place-america-n689141.
3. "Civilian Displacement: Hanford, WA," Atomic Heritage Foundation, July 21, 2017, www.atomicheritage.org/history/civilian-displacement-hanford-wa.
4. "Hanford Site: Hanford Overview," United States Department of Energy, updated January 8, 2012, archived May 11, 2012, web.archive.org/web/20120511135540/http://www.hanford.gov/page.cfm/HanfordOverview.
5. "RELEASES: The Green Run, Safe as Mother's Milk: The Hanford Project," accessed February 27, 2021, www.hanfordproject.com/greenrun.html.
6. John Stang, "Spike in Radioactivity a Setback for Hanford Cleanup," *Seattle Post-Intelligencer,* updated December 22, 2010, www

.seattlepi.com/local/article/Spike-in-radioactivity-a-setback-for-Hanford-916438.php.

7. Terry Richard, "Washington's Hanford Becomes Part of National Historical Park," *The Oregonian*, updated January 9, 2019, www.oregonlive.com/travel/index.ssf/2015/11/washingtons_hanford_becomes_pa.html.
8. Annette Cary, "Hanford, Animal Farm Advanced Radiation Research," *Tri-City Herald*, October 6, 2013, www.tri-cityherald.com/news/local/hanford/article32146353.html.
9. Annette Cary, "Workers Uncover Carcasses of Hanford Test Animals," *Seattle Post-Intelligencer*, updated March 31, 2011, www.seattlepi.com/local/article/Workers-uncover-carcasses-of-Hanford-test-animals-1225341.php.
10. Cary, "Hanford, Animal Farm Advanced Radiation Research."
11. Cary, "Workers Uncover Carcasses."
12. Robert Alvarez, "The Legacy of Hanford," *The Nation*, July 31, 2003, www.thenation.com/article/archive/legacy-hanford.
13. Annette Cary, "Hanford Contractor Awarded $7.8 Million in Annual Incentive Pay as Contract Is Expiring," *Tri-City Herald*, April 3, 2020, www.tri-cityherald.com/news/local/hanford/article241730461.html.
14. Annette Cary, "Hanford Cleanup Costs Triple. And That's the 'Best Case Scenario' in a New Report," *Tri-City Herald*, updated May 1, 2019, www.tri-cityherald.com/news/local/hanford/article225386510.html.

LEAP

MEERA SUBRAMANIAN

my entrails
dangle between paradise
and fear.

—JOY HARJO,
"The Creation Story"

1997 C.E.
Carbon dioxide level in the atmosphere: 362 ppm[1]

Oregon summers are for sleeping outside. By July, the legendary Pacific Northwest rains retreat and we have two months of skin-kissing temperatures and clear, star-studded skies. It's the season that makes suffering through the long gray winter worth it. After dinner we start a bonfire and surround it with our bodies and stories and songs. Soon it grows late and my boyfriend and I extricate ourselves from the group and find our way by headlamp to the bed we've set up at the edge of the forest. Nothing more than a tarp and a load of blankets piled on the ground just beyond the sagging back porch of our cabin, which is made of sculpted earth and century-old timber. There, in the liminal space between domesticated and wild, we enter the territory of our cat, Stranger.

I say that Stranger is "our" cat, but he isn't really ours. Nothing is. My boyfriend and I live in a community of non-blood relations on a forty-acre land trust, the deed held by a nonprofit. Stranger, a panther in miniature, belongs to the Douglas fir forest that rings the clearing where our community has cultivated a garden and built a cluster of cabins and a house of straw bales. Sometimes we'll spot his elusive form trailing behind us as we forge our way into the forest. Each time we enter the trees we take a different path, surrounded by trillium in the spring and chanterelle mushrooms in the fall. We move through the woods immersed, without fear. There are no poisonous snakes. No ticks. Maybe a brown recluse spider in the woodpile or bald-faced hornet nest under a log, but mostly we have to skirt just one threat: poison oak, its three shiny leaves waving like a warning flag.

Once we are sleeping outside, settled in our outdoor bed, Stranger hovers around us until his sleek dark body vanishes into the trees, shadowed from moonlight. He returns moments later with his prey of a small vole, and feasts beside our pillows. When he disappears again and returns with a bat, still alive, we leap up and wrangle it from him. We tuck the creature in a shoebox on a high shelf on the porch, out of Stranger's reach. In the morning, I awake at first light to the feel of the earth below me, sheltered by a canopy of fir needles. I have slept the sleep of gods, and the bat has flown.

2004 C.E.

Carbon dioxide level in the atmosphere: 377 ppm[2]

We can't stop fighting, and I'm finally the one to sever ties. The organism that was us dies. He and Stranger stay, and I leave, leaping into a new life that trades the Oregon woods for New York City, where the opportunities to become a professional storyteller expand, but the ones to garden naked evaporate. The only outdoor sleep I get is by sneaking up on a couple of rooftops, once to make out with a stranger who I shouldn't be kissing, and another time to *not* kiss a friend who I suspect would like to kiss me. Stargazing becomes a short-lived and atrophied activity, so I turn my attention to falcons, the aerial predators that live on skyscrapers and bridges. This is my wild way into an urban landscape, my attempt to write a new story that sustains me, to build a new world—solo—out of words instead of wood, in a place of concrete and steel. But something is seeping out of me, an energy force I can't identify.

700 B.C.E.

Carbon dioxide level in the atmosphere: 285 ppm[3]

More than a millennium before I move to what the Algonquian peoples perhaps called Manahatta, amid a cluster of islands in the Aegean Sea, people tell themselves stories of a different sort. They speak of Antaeus, a giant from Libya. The son of the sea god Poseidon and the earth goddess Gaia, Antaeus would accost passing strangers and challenge them to wrestling matches.[4] He'd always win. Every time. Until, one day, the demigod Heracles took him on and figured out his secret. He remembered the bloodlines that ran through the giant's veins. It wasn't his size that made him the champion, but the fact that every time Antaeus touched the earth, he tapped into his mother's might and his father's force, doubling his strength. That power could make men strong—when they remembered to draw upon it. It was the impetus that pushed mushrooms from the soil and flowers from their buds. It carried the salmon downriver into the sea and helped them find their way back home again to spawn and die. But Heracles figured out how to stem the flow of power. All he had to do to beat Antaeus was to separate him from the earth. He held Antaeus aloft, severed from the ground, and the giant's strength drained out of his suspended body like an ice cube melting in the sun. Heracles snapped him in two.

2007 C.E.

Carbon dioxide level in the atmosphere: 382 ppm[5]

I escape the city one fall day. Follow the curve of the Hudson River north into the Shawangunk Mountains to trail behind a biologist on his daily rounds. Find it curious as he shirks a field tall with drying grass, saying something about ticks. But our eyes are to the sky: the story I'm working on is about the falcons that inhabit the cliffs above us. That evening, I can't seem to find a decent campground and end up, along with a few others, paying fifteen dollars to pitch my tent in the backyard of some sketchy stranger. In the dingy shower, my soapy hand feels a bump on my butt and I recoil, yanking the beast out and flinging it down the drain. I later learn that this is an improper method of tick removal. There are stars outside my tent in the Shawangunks, and the earth below me, but my sleep is restless.

2011 C.E.

Carbon dioxide level in the atmosphere: 392 ppm[6]

I escape the city for good. I fall in love with a professor who's also a storyteller and he lures me to Cape Cod. Behind the cedar-shingled home we buy together, we smother a corner of the lawn with cardboard and have our new neighbor top it with a load of manure from his horses. I go exploring. Trails wind through conservation lands, through forests of oak and groves of holly. I learn to look out for poison ivy, similar to the poison oak of Oregon, both with their three-fingered wave. I follow a sign that says "Town Way to Water," tromping down an overgrown trail to the edge of a kettle pond. Delight in the discovery of the hidden water body. "I could make this place home," I think. That night, the delight vanishes when I find the brown abdomen and waving hind legs of *Ixodes scapularis*, the deer tick, protruding from my torso, its head embedded in my flesh, feasting on my blood.

I'm learning the rules of this new landscape. I extract the tick more carefully this time, cursing the creature. Cursing the probing, blood-sucking thirst of the arthropods that carry the viruses and bacteria that cause the illnesses that befall us. I am hearing too often about lives upended by Lyme disease. Reading too many front-page stories about ticks in the *Cape Cod Times*, along with stories of seaside houses tumbling into the ocean. I'm still writing about falcons, but also about water shortages and crop failures as far away as India, and something unsettling is setting in as I realize that it is all one story, and the story is changing how I relate to my world.

There are multiple reasons for the expansion of the tick's range. A warming world is one of them. That sky where I look for falcons holds more carbon dioxide than it used to, and thus more heat. The warmth stretches Cape Cod's summer like its famous saltwater taffy, leads to a longer season with fewer tick-killing freezes. On my way to the New Year's Day polar bear plunge in Barnstable Harbor, I pass a forsythia in bloom. In the summer, the rains come down heavier than ever. Ticks thrive on the moisture. They have more, as the scientists say, "reproductive success."

Scientists have also found that in the last decade and a half, tick-borne diseases have more than doubled in the United States and its territories. They account for more than three-quarters of all vector-borne diseases reported, and cases are known to be wildly *under*reported.[7] We humans aren't singled out by the bloodsuckers. Across New England and Canada, the moose are dying in droves. Imagine fifty thousand ticks feasting off one moose.[8] The great creature is brought down by blood loss before disease can even set in.

From Guatemala to the Sahel, Syria to Bangladesh, vector-borne diseases are thriving in a time of climate change, spreading among the people with the fewest resources to handle them. Sandflies with leishmaniasis. Lice with typhus. Tsetse flies with sleeping sickness. Fleas with plague. And the mosquito, carrying dengue, Zika, and chikungunya, a Kimakonde word meaning "to become contorted," as the victims curl into themselves with joint pain.[9] Several hundred million more people are at risk of malaria with a few more degrees Celsius, says the World Health

Organization, added to the nearly half million who already die each year.[10]

"Climate change," declares a governmental publication, has "far-reaching consequences and touches on all life-support systems."[11]

I think of Antaeus, reaching down to his life-support system, making contact, doubling his strength. And then I think of Heracles, severing the connection between that seemingly indomitable giant and the earth and snapping him in two.

2013 C.E.

Carbon dioxide level in the atmosphere: 400 ppm[12]

As far as we know, there hasn't been this much carbon dioxide in the atmosphere for three million years.

I'm not thinking of this when, on a tire swing under a load of stars in his parents' backyard, just ten minutes from our home, the storyteller asks me to marry him. We seal it when he slips a clunky Ace Hardware ninety-nine-cent lock washer over my finger. I am joyous. So are his parents, but they're also tired. In their late seventies, they're both recovering from babesiosis, the latest tick-borne disease to arrive on the Cape.[13] Their diagnoses are 2 of only 1,762 cases this year across the entire country.[14] Lyme still dominates, but the list of other pathogens that lurk in tick spit has grown to seven on the Cape.[15] In addition to Lyme and babesiosis, there's anaplasmosis, Powassan, Rocky Mountain spotted fever, Heartland virus.[16] New types of ticks have shown up, too. Along with deer ticks and dog ticks come lone star ticks. They all show up here in New England, where a power plant is named after the pilgrims who once landed here, fracturing the lives of the Mashpee Wampanoag with diseases the newcomers carried. The names of the ticks and the diseases they harbor reveal their displacement.

I feel less displaced, more at home here, but new and disconcerting habits I didn't have in the late 1990s have set in. I find myself pausing at the edge of the forest. To step into the woods, to encounter any type of brush, to expose my midriff because of an untucked shirt, or an ankle from a slouching sock, is to become vulnerable. I stop looking for the mushrooms or cascading down to the kettle pond sunk in the hollow. Exploring the path isn't worth an encounter with what might be lurking there.

We develop a new ritual in early summer. Instead of setting up an outdoor bed, we decide which will be our set of outdoor clothes for the season. I clip them to the clothesline and douse them with permethrin, an insecticide that mimics the repellent quality of chrysanthemum flowers, until my wrist hurts from squeezing the spray bottle trigger. Now, when my love and I go for a walk, we tuck our pants into white socks, don hats to cover our heads, and basically prepare for battle.

2017 C.E.
Carbon dioxide level in the atmosphere: 406 ppm
I've nearly lost touch with my West Coast ex, but word comes from other friends that Stranger no longer emerges from the forest; he is presumed dead. There's also news that Oregon is aflame with a thousand fires. As the ticks have spread, so too have the wildfires, becoming more sudden, more intense.

Summer temperatures have transformed into scorching ones, the fire season extending year-round in some places. A friend along Hood River tells me her plans to go hiking were canceled, again, the air choked in hazardous black smoke. On the Cape, I deftly remove a tick from the scalp of my friend's young boy, using a tick-extracting device I carry on my key chain, always at the ready.

2018 C.E.
Carbon dioxide level in the atmosphere: 408 ppm

I keep telling myself that it's the ticks that have changed my relationship to nature. With each story I hear from my scattered loved ones, I realize that it's a lie. The friend who lost his Red Hook apartment to Hurricane Sandy, the Indian aunt stranded in her Chennai apartment after the flood, the friend in California who fled the fires. With each new reported story for my journalism—the failed peach harvest because of the too-warm winter, the dogsledders in Wisconsin who can't depend on the snow, the North Dakota rancher who sold off her stock during the drought—I experience the lie anew.

It isn't about the ticks. They are just living their intrinsic ticky lives, doing the only thing they know how to do. They hover on a stem, forelegs open wide, reaching with desire and hunger, waiting for the exhaled breath of carbon dioxide that signals a mammal passing by, and then—leap. They probe their way in, and drink. Biological life forces lead the way, driving the actions of all beings, a tick, a coronavirus, you, me.

How I relate to the ticks has become emblematic of my

entire relationship with the natural world, that wild and rank place I used to immerse myself in completely. It restored me. I could disappear into the Oregon woods and come back calmer. I could lie at the base of a tree amid the leaf litter without fear and let my gaze follow the trunk up to infinity, inducing vertigo, and feel like I was flying. I didn't worry about ticks in New York City either, could stand on the Brooklyn Bridge with binoculars and a contraband bottle of wine looking for falcons and feel the same brilliant giddiness. I could forget everything in those places and move without fear. I could sink in and lose myself. Find myself.

Now I can't stop the calculus in my head as I interact with the places that once offered solace. This is what climate change is. It's what it does to the psyche, along with the body and the places we love. It's nearly invisible until the moment something startles you into attention. A creeping catastrophe, waiting with arms outstretched to deliver a suffocating embrace. And once the knowledge is gained, there is no unknowing it. You are no longer climate blind. You see and cannot unsee.

We used to be a story in nature. Now we are *the* story. There was a time when *Homo sapiens* used narrative to explain the inexplicable. Gods of the sea and goddesses of the earth. Then some of us forgot the stories. Some of us cut paths through the territory of the unknown with increasingly bigger machines, thinking we'd gained control of our world. That was a lie, too. We are back in a time when the land again acts in disorienting, incomprehensible ways. Ten thousand years of living in a steady climate is over. We have returned to the times of mythology, and we need new stories to survive.

2020 C.E.

Carbon dioxide level in the atmosphere: 416 ppm[17]

With each flip of the calendar, as I labor at this draft, I have to update the figure that marks the carbon dioxide level of our atmosphere. By March I'm spraying down my overalls with permethrin but I've nearly forgotten about the ticks and their disabling bacteria, as a virus rewrites our world this spring. This wasn't the new story I was expecting. Or any of us, really—except for the epidemiologists and the disease experts who've been warning us for years. We disregarded them as we have the climate scientists who've been cautioning us for decades.

But some are listening. Some are already writing the new stories we'll need to survive. I sought them out over the years, as climate facts turned into knowledge, as the far-off future became the present. Found them among young West Texas wind turbine technicians, apolitical and aloft in their power offices that spin in the sky. Farmers in Rajasthan growing soil as much as crops. Tribes of women who gather in kitchens for squash soup as they envision new ways of being in the world, and with each other. Young evangelicals fighting to save God's creation, and the Lakota Water Protectors who remind us that water, not oil, is life. For now, as I and others stay home to stay safe, that carbon dioxide level takes a dip. I tend the garden and turn to sci-fi writers who imagine the best, and worst, of what we can become. In real life, half of Main Street has been taken from the cars and granted back to pedestrians. New stories are emerging.

Ekstasis is the Ancient Greek word for that moment of dislocation when we step out of ourselves, become unbound, when something beyond our bodies engages us to the point that we transform. Will climate change force this upon us? Could we take it as an opportunity to decide what that future could look like, not trapped but ready to leap?

We could be the next generation of Poseidon and Gaia's love children. Antaeus almost had it right, but not quite. The giant squandered his superpower. He touched the earth to redouble his strength and then wasted it accosting strangers. We could draw on the same earth force, but instead of using it to fight each other, use it to refill our pens and recharge our minds and repower our worlds with energy spun from light and wind. Wander through the woods, soles to the soil, and emerge transformed, our anger and terror turned into something revolutionary. The mythologies of future generations are ours for the making. What story shall we tell?

Notes

1. Rebecca Lindsey, "Climate Change: Atmospheric Carbon Dioxide," NOAA Climate.gov, August 14, 2020, www.climate.gov/news-features/understanding-climate/climate-change-atmospheric-carbon-dioxide.
2. Lindsey, "Climate Change."
3. Dieter Lüthi et al., "High-Resolution Carbon Dioxide Concentration Record 650,000–800,000 Years Before Present," *Nature* 453 (2008): 379–82.
4. "Antaeus," GreekMythology.com, accessed February 27, 2021, www.greekmythology.com/Myths/Gigantes/Antaeus/antaeus.html.

5. Lindsey, "Climate Change."
6. Ibid.
7. Ronald Rosenberg et al., "Vital Signs: Trends in Reported Vector-borne Disease Cases—United States and Territories, 2004–2016," *Morbidity and Mortality Weekly Report* 67, no. 17 (May 4, 2018): 496–501.
8. Laura Poppick, "As Winters Warm, Blood-Sucking Ticks Drain Moose Dry," *Scientific American*, December 11, 2018, www.scientificamerican.com/article/as-winters-warm-blood-sucking-ticks-drain-moose-dry; Chris Bosak, "For NH Moose, Winter Tick Can Be a Deadly Mismatch," *Keene Sentinel*, updated August 28, 2019, www.sentinelsource.com/news/local/for-nh-moose-winter-tick-can-be-a-deadly-mismatch/article_cfc8fae3-617b-57bb-bc77-60225616a184.html.
9. "Vector-Borne Diseases," World Health Organization, updated March 2, 2020, www.who.int/news-room/fact-sheets/detail/vector-borne-diseases; "Chikungunya," World Health Organization, updated September 15, 2020, www.who.int/news-room/fact-sheets/detail/chikungunya.
10. Carolyn Beeler, "Climate Change Will Make Animal-Borne Diseases More Challenging to Predict," *The World*, February 14, 2020, www.pri.org/stories/2020-02-14/climate-change-will-make-animal-borne-diseases-more-challenging-predict; "Malaria," World Health Organization, updated November 30, 2020, www.who.int/news-room/fact-sheets/detail/malaria.
11. A. K. Githeko et al., "Climate Change and Vector-Borne Diseases: A Regional Analysis," *Bulletin of the World Health Organization* 78, no. 9 (2000): 1136–47.
12. Robert Kunzig, "Climate Milestone: Earth's CO2 Level Passes 400 ppm," *National Geographic*, May 12, 2013, www.nationalgeographic.com/pages/article/130510-earth-co2-milestone-400-ppm.

13. "Babesiosis—Data & Statistics," Centers for Disease Control and Prevention, updated January 13, 2015, www.cdc.gov/parasites/babesiosis/data-statistics/2013.html.
14. Ibid.
15. Rosenberg et al., "Vital Signs."
16. Ibid.
17. "Daily CO2," CO2.earth, accessed March 16, 2020, www.co2.earth/daily-co2.

COME HELL

LACY M. JOHNSON

It started raining in the fall and kept raining even as the temperature dropped, when the rain became snow, falling like scraps of wet cloth, which collected in giant drifts. In the spring, the snow turned back to rain and kept falling, and by summer, the creeks and rivers had filled and left their banks, submerging first the low farmlands, then the farmhouses and barns and steel outbuildings, and then also the train tracks, abandoned coal mines, the highways and bridges, so that an archipelago of grass and silos and trees stretched between the Missouri and Mississippi Rivers from South Dakota to the Gateway Arch. The rain kept falling even through the summer, when the rivers, grown to monstrous size, approached the edges of towns and then filled the streets. People moved their homes and businesses—moved entire towns—to higher ground. In the summer, when the river was at its widest and deepest, the days underwater accumulated: seventy-five, ninety-seven, one hundred fifty. This was Missouri in 1993.

A year before the rain started, my parents had sold the farm where I'd spent what felt like my entire childhood and moved into town to a house only five blocks from our church. Living so close to the pulpit meant that not even the worst flood in sixty-five years could keep us from attending service every Sunday morning and every Wednesday night. As thunder shook the stained glass in the sanctuary, our pastor, Brother Michael we'll call him, told us, his faithful congregation, how God had sent the flood of the Old Testament to cleanse the Earth of sin and moral depravity, how he had instructed Noah to build the Ark and gather the animals, and how this flood, our flood, was also an instruction from God to gather to one another and to Him, to "be fruitful and increase in number and fill the earth." God made the Earth for us, Brother Michael told us from the pulpit. He made the sun for us and the rain for us, even rain that falls for an entire year. My family sat, as usual, near the rear of the church; my father nodded as Brother Michael raised his voice and his hands. The rain will cease and the clouds will part, he told us, and the rainbow that God will bend down from Heaven will be the token of that everlasting covenant between us and Him: "Just as I gave you the green plants, I now give you everything."

God had given me as many summers on the farm as I could remember, and now that we'd moved to town I could hardly find a thing to do. Our farm hadn't even been very big—only seventy-six acres, and much of that was forested—but I could spend the whole day walking the perimeter and learn the history of the weather in the shapes of the land: the way the creek bottom held a millennium of flooding in its black soil, the way trees on the ridge that marked the edge of our property had died and toppled years ago in a drought. But our new house sat right next to our neighbors on both sides, and life felt enclosed and small. I could ride my bike across the entire town in less than a morning and see only ghosts of what the town had once been: an abandoned coal elevator behind the Benjamin Franklin downtown, the sharp bare ridges of surface mines that sloped into deep black pits, the rail lines that once carried coal from the local mines to the power plant where my father worked, and now carried it to the plant from much, much farther away. When it was raining, I'd sit on the floor of my upstairs bedroom and try to record songs from the radio onto cassette tapes while my sister read the funny pages from the newspaper in the other room, the headlines announcing the river cresting, another record, thousands losing access to drinking water, to electricity. Sometimes my sister and I would walk down the street to the edge of the water to throw leaves and sticks and watch them river away.

During the worst of the flooding, when my father would leave the house every morning to drive his pickup truck along the dry backroads on his way to the power plant, my mother would cross her arms and let out a long, tired sigh. I'd been to the power plant maybe only one time, but I could hold an image of it clear in my mind: the green fields cradling a deep blue lake that stretched for miles, surrounded by a forest of maple trees. At the southern tip of the lake the land was charred gray and black—no trees, no grass, no flowers or even any weeds—and in the middle of this burnt-out place, right against the edge of the lake, three churning smokestacks belched black smoke that on a clear day I could see from miles away.

I didn't understand what my father did for work at the time, but I knew the power plant provided electricity to dozens of towns' worth of stop lights, attic fans, refrigerators, and single lightbulbs swinging from wires in basements. I also knew my father had been raised on a farm, like me, and on the farm he had studied the gospel of the planting and the harvest, the cycle of hunting in the winter and swimming in the summer, and of watching fog roll down from the hills when summer turned toward fall. He understood what it meant to bale hay and birth calves, and also that he didn't want to do that sweaty, back-breaking labor his whole life if he didn't have to, so he went to college and learned how to turn coal into electricity and wear a collared shirt. Now that he worked at the power plant, my father believed what Brother Michael told him in church: that coal is a blessing if we use it to power our cities; that oil is a blessing when we turn it to gasoline. God gave humans the Earth, my father believed, and we please God when we use His gifts to fulfill our needs.

I learned only later that coal had once been abundant in the part of Missouri where we lived—fields of it as wide as entire counties; seams of it tatted under the land like black lace—but when the coal mines started to run dry, and mining local coal got more expensive than shipping it in, the mines closed, the miners left, the town withered, and new coal, "cleaner" coal, started arriving at the power plant by the truckload to be spray-cleaned, cracked, crushed, pulverized, and burned to produce the dirtiest kind of electricity—each ton of coal producing almost two tons of carbon dioxide, a billion tons of coal burned in the United States every year, polluting the air we try to breathe as much as all of our cars, trucks, buses, and planes combined, enough to melt glaciers in the Himalayas and the ice caps on the planet's two poles, enough to raise the sea level of every ocean on Earth, and to cause increasingly severe disasters every year—decades-long droughts that turn farmland to desert, wildfires that turn ancestral forests to ash. I didn't know whether my father knew that then. I didn't know that then.

All I knew was that when my father came home from work every evening, we'd gather at the dinner table and he would ask us to bow our heads to pray. Some nights we'd thank God for the food on our table and the clothes on our backs; other nights we'd ask God to help the men who'd lost their farms in the flood, who'd lost their jobs at the power plant, who'd lost their homes, who'd lost everything. After my sister and I cleaned up all the dishes and loaded the dishwasher, my father would fall asleep in his armchair while the rain filled the streets and the news played on television—the living room flickering with footage of the swollen rivers, brown and fast and so wide in some places that I held my breath while the camera tried to take it all in: cars underwater, crops underwater, women walking through water carrying baskets, men carrying their children on their backs, houses bobbing in the water, grain silos underwater. The weatherman ran his hands through his hair while water rushed through broken levees. Baseball stadiums filled with water; American flags flapped from submerged poles; men stacked sandbags outside their storefronts; coffins bobbed in shallow lakes—semi trucks underwater, gas stations, train cars carrying coal to the power plant where my father worked. The coal he burned at the power plant had no more to do with the flood that was drowning every farm from here to South Dakota, my father believed, than the way he tied his shoes.

I never once heard my father say the phrase "global warming" or "climate change"—not until decades later, when he had already decided he didn't believe they exist—though I often heard him say "Heavenly Father" and "Amen." One night in late summer, when the rain had finally started to let up, my father had just finished blessing our dinner when the phone rang in the kitchen. My father stood up, folded his napkin onto his plate, and walked to where the phone hung from the wall. He lifted the receiver and listened for a moment before holding it out to me. I pulled the long spiral cord into the bathroom and a friend on the other end whispered that there was a party later, out where all the roads were flooded, out past our old farm, where the sheriffs couldn't reach us in their low Crown Victorias. I didn't hesitate to lie to my parents about where I would be. I rode in the backseat of a tiny hatchback packed with more teenagers than seat belts down a long gravel road to the water's edge, then climbed with the others into the bed of a tall pickup truck and sat on the wheel well and held tight to the edge. The driver eased us forward into the water—a dark, glistening mirror for the cloudless night—and ferried us slowly to the dry ground on the other side, to wet hay bales and a fire that burned while my classmates passed bottles and bottles of cheap alcohol. Hours later, I lay back in the wet grass and looked up into a sky of stars spinning above me and though I wanted to feel God in everything, to feel certain of Him like my father did, all I felt was the movement of the water all around us, carrying us toward a future I couldn't yet see. The gospel I learned from the forests and the trees, the creeks and the water, the crops and the land, I decided, was that the Earth belonged to itself; none of it belonged to me.

After the water receded, and the farmland dried, and the old pit mines drained, and the farmers began tilling the river sand deep into the fields that produce less and less every year, my father left for work each morning, driving to the power plant on the highway instead of the back roads, and the three smokestacks went on belching black smoke day after day, year after year, like nothing had changed. We went to church every Sunday morning and every Wednesday night, and before dinner my father asked us to bow our heads to pray. For a while, my sister read the funny page from the newspaper, and I sat on the floor of my bedroom recording songs from the radio onto cassette tapes while trying to imagine what I would take with me when I left. What I loved most about that place was the warm black soil on the creek bottom, the fog rolling down from the ridge of trees, the way the wind rippled the fields of grass in waves. But none of those things were mine to take, and besides, by then they were already gone.

AFTER THE STORM

MARY ANNAÏSE HEGLAR

"Granddaddy! GET BACK IN THE HOUSE!"

Of all the things I thought I'd be doing on this visit back to Mississippi, yelling at my grandfather in the middle of a hurricane wasn't one of them. I was home for what I thought would be a one-week vacation between a summer in New York City and my senior year at Oberlin College.

I never thought I'd yell at my grandfather, ever. He was my grandfather, we are Black, and I like having teeth in my mouth. My grandfather never raised a hand to me, but I just assumed that any sort of backtalk would release a giant rock from the sky to smite me.

On the other hand, I never thought I would see a hurricane in Port Gibson, Mississippi, either. We're no stranger to thunderstorms, floods, tornadoes. But hurricanes? That's a coastal problem. The "port" in Port Gibson denotes its position on the Mississippi River. We are about two hundred miles from the Gulf Coast. But Katrina went where she wanted.

Maybe that was why my grandfather thought it was a good

idea to recover the feeder for his beloved hummingbirds after the wind knocked it down. It was all so unbelievable, so why believe it?

"Granddaddy." I tried to soften my voice. "It's a hurricane. The birds aren't out right now."

"What do you know?" he shot back. "You not a bird."

I couldn't argue with that.

But I didn't have to. As soon as he got off the back porch, Katrina declared her dominance and knocked him off balance. A man for whom confidence was everything lost it all to the wind. He came shuffling back toward the house, avoiding the concern in his granddaughter's eyes.

My grandfather was a very proud man. I don't think I'd ever seen him lower his head or shrink his shoulders. As a Black man who grew up in Alabama in the 1920s and 1930s, served in the military in the 1940s, integrated the schools in Nashville with his own children in the 1950s—he had a lot to be proud of. He passed a lot of that down to me, almost by osmosis. He didn't talk about it much, but I could feel it in his presence. Something about being near him made you want to stand straighter and speak clearer. Ever since I could remember, I was terrified of disappointing him, and desperate to impress him. It wasn't easy.

Now he said nothing. He just stumbled back into the house, where my mother had cable news pundits and meteorologists blaring in every room.

Things hadn't gotten bad yet. The power was still on. The water was still running. And I was in the middle of an ill-advised experiment of steaming okra. I would never try that again.

We were worried for New Orleans, that beautiful, beautiful

city in a soup bowl. Our regional jewel. But we also felt relief because, that morning, it had been announced that Katrina had not hit New Orleans head-on and had instead made landfall at Bay St. Louis, Mississippi. Of course, we were worried for the people there, but at least, we thought, the loss of life would be contained.

There was a lot I didn't know about then. But years later, I would see that the eye of this storm was forming the lens through which I still see the climate crisis today: one in which structural racism and inequality collide with fearsome extreme weather to reveal the grotesque unnaturalness of disaster.

Because Katrina's aftermath was so horrific, we forget how utterly strange she was as a storm. We forget that she made landfall in Florida as a meager Category 1 hurricane before sweeping back out to sea to gather more strength for the Gulf Coast.

We forget that, by the time she made landfall, she had weakened from a Category 5 to a Category 3. But what Katrina sacrificed in strength, she more than made up for in size. At the time, she was the largest hurricane ever to hit the United States, affecting millions of people over approximately ninety thousand square miles. And that was just in the short term. Just before our electricity went out for what would become a week, we saw that Katrina was covering the entire state of Mississippi. From the Coast to the Delta.

We forget the tornado outbreak she spawned as she traveled over land. Fifty-seven tornadoes over the space of eight states—from central Mississippi to Pennsylvania. With eighteen tornadoes across Georgia in a single day, she far exceeded the state's previous daily tornado record of two.

The other thing often left out of the narrative, but which I can never forget, was that Katrina descended the day after the fiftieth anniversary of the murder of Emmett Till. If you are Black, and especially if you grew up in the South, the name Emmett Till brings immediate, arresting, gruesome images to mind. The name sinks to the bottom of your stomach like a bag of rocks—or like the cotton gin fan that forced his barely pubescent body to surrender to the Tallahatchie River.

The anniversary was the biggest news story in Mississippi in the weeks before the storm. How far had Mississippi come? Had we stood still? What comes next? There, in my three-generational home of Black southerners, I couldn't *not* think about the anniversary, even then, with the storm overhead.

I remembered the meteorologists explaining how hurricanes start off the coast of Africa and gather strength as they cross the Atlantic, following almost exactly the route of slave ships.

I wondered if this Category 3 storm was really a fourteen-year-old boy named Emmett.

If I remember correctly, there were no casualties in Port Gibson and the property damage was minimal. Almost everyone got roof damage, and lots of yards had fallen trees—but nothing that couldn't have happened in an exceptionally strong thunderstorm. We were lucky and we knew it. But still, we held our breath for news about the coasts and New Orleans.

Our only source of news was our battery-operated radio. The local NPR segments dripped with such overt racism, it was impossible to trust them. You could hear it in the way they described certain neighborhoods and the people who lived there. "They" were looting. "They" were rowdy. The armed vigilante groups had no choice but to defend themselves, the radio hosts said. But we knew those neighborhoods. We heard those dog whistles. These communities had just been devastated to an unimaginable degree. Who wouldn't be rowdy? Who wouldn't take what they needed from a store that would be closed indefinitely? Were they supposed to wait for help from above?

We waited all day, instead, for the national broadcasts. That was the only thing that kept us even halfway informed about the unfolding tragedy. Here we heard people described as, well, people. We heard stories of loss and angst from all over the city, in every neighborhood. We heard about people in need, people waiting for help that wasn't coming, so, yes, they broke into a store. We heard about rumors of violence, but on these national broadcasts, they were also described as what they were: unconfirmed rumors.

(The woman who did that reporting is now my colleague at NRDC. I've thanked her over and over, but I still don't think she knows what she meant to us.)

We lost water for a little less than a week. That was the hardest part, because of how hot it was right after Katrina. I don't remember the exact temperature, but that day is seared into my memory as the hottest I've ever lived through. I know it was over a hundred degrees because that was as high as our back porch thermometer could go. I tried to curb my water intake, to save it for my grandfather and my mother. We all slept a lot, including the dog.

My mother's car was in the shop, so we couldn't even go for a drive to cool down in the car's air-conditioning. That really became a problem once the groceries started to run low. At one point a neighbor let us tag along on a trip to Vicksburg to go to Kroger. Later, I stubbornly walked to the grocery store in town and was apprehended, scolded, and driven home by another neighbor.

I was supposed to leave to go back to Oberlin a day or two after the storm, but I had to put that off for more than another week. I'd flown into Dallas and taken a bus to Mississippi, and planned to do the same to go back, but the roads to Texas were littered with fallen pine trees. There was literally no path to Dallas.

We went without power for about a week, and without phones (both cell and landline) for two or three days. We were essentially cut off from the rest of the world, but Mississippians are no strangers to blackouts. Blackouts are part of life there. You expect them. They force you to hold still, to be patient. Especially at nighttime, when the fever of day breaks and the crickets and frogs play their symphony. You can close your eyes and find the beauty in being exactly where you are.

We never knew when a tornado or a thunderstorm would knock power out, or when a tree would fall on a power line, or when the grid would simply get overwhelmed. Since we couldn't predict it, we just stayed ready. Everyone had flashlights and batteries and candles, and most people had battery-operated radios. Years ago, I'd grown tired of calling my mother during bad thunderstorms and tornadoes only to get a dial tone that fed my panic. So I made her buy a battery-operated phone. The town hospital around the corner had generators. That's where I went to charge my cell phone when it was finally working again. The elderly woman who lived next door also had a generator for her breathing machine.

The landline came back before my cell phone did, but allowed incoming calls only. We fielded call after call from distraught family members. Each one gasped when they finally heard my mother's voice or mine. They'd been calling for days.

Most of my relatives live in Birmingham or Houston or Atlanta or Washington, D.C., and had never been to Mississippi. They didn't know if the newscasts were exaggerated. We didn't know ourselves how bad the damage was until we talked to my brother, who told us the interstate had been broken up like dominoes. That's when it became real.

When my cell phone came back on, it was full of increasingly distressed voice mails from friends at Oberlin. Apparently, there had been an automatic message that stated, matter-of-factly, "Due to the hurricane in the area you are calling . . ." They didn't know how close I was to the coast or how much danger I was in. And because these were the days before text messages, they had no recourse other than to leave voicemail on top of panicked voice mail.

When the power came back on, we saw everything with our own eyes. We saw that the towns on the coast had been completely washed away. I can still hear Governor Barbour's voice: "I don't mean they were badly damaged. I mean they're simply not there."

We saw beautiful, beautiful New Orleans flooded to her brim. We saw pictures of the vigilante groups that patrolled white neighborhoods to keep Black people out. Again, I thought of Emmett and his open casket, as I watched New Orleans and the coasts turn into open graves.

We saw the "looters" and heard one of them shout to the camera in that beautiful, melodious New Orleans accent, "Yes, we stole the shoes 'cause all ours got lost in the storm!"

We saw the overhead footage of all the people stranded on their roofs. It stretched so far it defied any semblance of a border. We heard conflicting reports from channel to channel, segment to segment, about violence in the Superdome, in the Convention Center, on the Danziger Bridge. Reporters described people shooting at police helicopters from their roofs, but also of people so desperate for help that they shot into the sky to signal distress.

I thought about how hot those people must have been. We were suffering with no fans or air-conditioning. They were suffering under the direct glare of the sun. Children, pregnant women, elderly people. The swamp reclaimed the city. Snakes and alligators and fish swam in equal terror through swallowed neighborhoods, only the roofs peeking out from the gray water.

Growing up in the Mississippi River region meant growing up in both the shadow and the embrace of New Orleans. We had Mardi Gras parades, and it was easy enough to find King Cake. It wasn't unusual to see ATMs with French as a language option.

If the day was clear enough, we could point the antenna just right and hear the radio stations from Baton Rouge that played the newest Master P, Hot Boys, and DJ Jubilee before we heard them anywhere else.

It broke my heart to see these people, whom I'd always known to be as generous with their culture as they are with their laughter, suffer so hideously. We'd always known that New Orleans was unlike any other place in the country, or the world, but we never thought we'd see New Orleanians referred to as refugees in their own country. It was as devastating as it was unbelievable.

I never thought that I'd see the Mississippi my grandfather had known when he was my age, or even the one my mother saw. The Mississippi that brutally murdered a fourteen-year old boy for a wolf whistle that we now know never happened. But Katrina revealed fault lines that I could never unsee.

Those images would haunt me forever, and they still frame the way I look at the climate crisis in my work today. Thanks to Katrina, I can't look at the climate crisis without seeing the grimy fingerprints of slavery and Jim Crow and colonialism and genocide and patriarchy. It's what happens when large swaths of people are not only systematically "left out," but forced to be their own gravediggers and pallbearers. Now I can't help but see who is saved and who is abandoned. Whose bodies litter the road to the "greater good." And how none of it is an accident.

After that summer, I never saw my grandfather the same way. He'd already begun to show symptoms of dementia. In the years after the storm, I saw him become less cognizant, less there, with every visit home. We lost him in 2012.

I never saw New Orleans the same way, either. The next time I visited was about ten years later, and the grit of the storm was still there, on every billboard, every building, every face. There was construction everywhere—not to build, but to rebuild. Homes were still boarded up, with giant orange *X*s painted on the outside next to marks that tallied how many bodies had been found inside. To this day, everything there is dated as either "before the storm" or "after the storm"—and no one questions which storm.

Like my grandfather, New Orleans became more fragile, more tenuous. I saw the things that made them both—the pressure that made the pearl—in a way that I never had before. They became more beautiful, more precious. And I couldn't unsee it.

WALKING ON WATER

RACHEL RIEDERER

I stared out the window of the Jeep at the spot where the Nile had been halved. One side of the riverbed had been completely filled in—a combination of soil, stone, and concrete piled high and dense enough to force all of the water to the other side. In the driver's seat of the Jeep was Kenneth Kaheru, a civil engineer and manager at the Bujagali Dam construction site just outside the town of Jinja, Uganda. It was July 2009; two years later, the transformation would be complete, and in place of that expanse of earth and river would lie the foundation of a massive hydropower dam. Dams on this scale have to be built in two parts, Kaheru explained to me, so his company had filled in one side of the river first. They would build it up, then open a spillway to extend the structure to the other side of the riverbed.

The construction site was downstream from another dam, one completed fifty years earlier by the British colonial government. That older dam marks the edge of Lake Victoria, the largest lake in Africa. Water spills through the dam's turbines, creating a big part of the nation's electricity, before emerging to form the headwaters of the White Nile. It wasn't far from the guesthouse where I was staying, a little house run by a group of siblings in their teens and twenties, whose parents lived in one of the villages outside of town, and populated by a bunch of British teens on a gap year, and me. I'd visited the dam as soon as I'd arrived in town. Looking over the edge, I hadn't been able to comprehend how anything man-made could hold back such a large body of water. I didn't have time to ponder the view for long—a soldier with a rifle approached and told me to keep moving—because of its role in the nation's energy production, the dam is a national-security asset as well as a wonder of physics.

I had come to East Africa to write about a trio of dramatic environmental issues that were plaguing Lake Victoria. The gigantic Nile perch, an invasive species of fish imported during the colonial era, had gobbled most of the lake's other inhabitants to near extinction, turning the lake into a giant monoculture. Another invasive species, the water hyacinth, an ornamental plant with long purple flowers that grows in thick mats on the surface of the water, had become so overgrown that it had, in some places, choked out whole docks and ports—scientists and local fishermen were beating it back with everything from machetes to dynamite. But to me, the most intriguing of these problems was the question of the missing water.

For years, Lake Victoria's water level had been dropping drastically, and there wasn't a consensus about how or why. A mild drought had recently plagued the lake, and some population growth in the region had increased the amount of water use, but neither change fully accounted for the way the water was slipping away from the waterline it had held for decades. But while traveling around the lake pursuing these stories, I'd gotten sidetracked by the new dam. The Bujagali Dam was being hailed by the Ugandan government and the World Bank as a source of clean and renewable energy. Its sheer scope was arresting: when completed, it would control the headwaters of the Nile.

It was also the center of a heated controversy—many big dams are. They displace people and disrupt ecosystems. They are complicated from a carbon perspective—certainly cleaner than burning fossil fuels, but their reservoirs are sources of surprisingly high emissions of the global-warming gas methane. And, as John McPhee writes in his 1971 *Encounters with the Archdruid*, dams are rich with symbolism. "The conservation movement is a mystical and religious force, and possibly the reaction to dams is so violent because rivers are the ultimate metaphors of existence, and dams destroy rivers," he writes. "Humiliating nature, a dam is evil."

I was not yet a student of dams, or the movements against them, but it was clear even in my naivete that the Bujagali project embodied especially weighty tensions. The nation needed energy, and building a new hydropower dam was one way to get it. But the dam would also displace the indigenous Basoga community that lived along the river, and further throw off the water balance of the lake that millions of people depended on. So often, the questions of environmental justice are subtle, played out in particulate-matter levels invisible to the naked eye or across time scales too slow to register as emergencies until it's too late. But with this dam, central questions of environment and development—How much is enough? Which communities have to bear the costs? and, crucially, Who gets to decide?—were being made visible in real time. Their answers, in this case, would emerge from the building of a massive structure extending nearly 100 feet into the ground and 150 feet above. The project was divisive, though, for reasons beyond the usual: as one local professor explained to me, "For the Basoga, the Bujagali Dam will mean spiritual death."

Touring the Bujagali construction site, with its mountains of gravel, battalions of earthmovers, even buildings to house on-site cement production and stone-crushing—I could see a controlled fire burning in one such structure—I began to understand how these hydrologic leviathans come into being: through the accumulation and exertion of a huge amount of material, machinery, energy, capital, and will. I was fascinated with the dam—and with all of Lake Victoria's environmental troubles—because it seemed to make clear the connections and interdependence among vast systems that were masked in my own experience. Living in New York City, I was separated from the sources of my own food, water, and power by miles, by complex arrangements of shipping and purification and transport and infrastructure that I depended upon but that were all invisible to me. That trip taught me to investigate these systems in my own life; at the time, I just wanted to go to a place where they were more on the surface, more easily seen.

At the Bujagali construction site, the banks of the Nile were steep. There, water pushed against the riverbed, smashed into the partially submerged boulders where cormorants perch and dive. It frothed against the few small islands covered with primordial-looking ferns that had managed through some tenacious geology not to be eroded or carried downstream wholesale. The riverbed's downward gradient wasn't dramatic—the water tumbles forward so fast it's easy to overlook that it's also falling downhill—but it was enough to give the water tremendous force and speed, pushing even against its own surface, mixing with the air above it. The water that had looked like an expanse of black glass, during its rest in Lake Victoria, became an opaque white froth.

Today, the water no longer looks like that. The dam that was being built—and contested—when I visited in 2009, was completed in 2012. The Bujagali Falls are now submerged beneath the surface of another reservoir.

This spiritual controversy had to do with the falls from which the then-embryonic dam got its name. The Bujagali Falls did not look like my conception of a waterfall—rather than a cascade dropping in a vertical wall, they were a series of descending whitewater rapids. They were just a few kilometers upriver from the construction site, in the zone that was submerged when the dam was completed. I kept thinking about what the professor had told me about "spiritual death," and the vague references I'd kept hearing about how the falls were "spiritually significant" to the Basoga—but I was nagged by the distant language, and wanted to understand what it meant for a stretch of rapids to be so holy.

Since the dam was funded by the World Bank, it had to follow the bank's rules about protecting culturally sacred places, which dictate that the funder and developer must work with community leaders to mitigate irreversible cultural damage. In the case of Bujagali, that meant working closely with the Basoga's local spiritual leader, the Nabamba Bujagali, for whom the falls were a sacred site, a home to both gods and ancestral spirits. To prepare for the construction of the dam, the hydro developer, AES Nile Power, needed to engage the spiritual leader to relocate the spirits from the falls. That leader, a type of shaman, was to move them—via a spiritual ceremony—to some nearby shrines, where people could continue to worship or pray after the relocation.

The problem was that at the time, it wasn't clear—at least to outsiders—who the Nabamba Bujagali was, because there were two men claiming the title. The older of the two went only by Jaja Nabamba Bujagali; he didn't use—and nobody else seemed to remember—the name he had before he'd adopted the religious title. The younger man also claimed the title of Nabamba but still went by his original name, Benedicto Nfuudu. (To avoid confusion, I will call the older man Jaja and the younger, Nfuudu.) Years before, AES Nile Power had contracted with both men to perform the ceremonies and get the falls spiritually ready for submersion, a sort of hydrological last rites. This didn't go as the developers had hoped.

It's hard to know exactly what happened at those spirit-relocation ceremonies, which took place in the late 1990s. I've read accounts by foreign correspondents from the United States and Europe that include a bonfire, dancing, and the sacrificing of a chicken and a goat. Other retellings are more sedate. A member of the World Bank Inspection Panel sent to assess the situation in 2001 wrote: "Both mediums have contacted the spirits on behalf of AESNP. Both have signed contracts with AESNP." Nobody I met during my time in town believed in the authenticity of both Jaja and Nfuudu—their relationship seemed strictly either/or. But the World Bank and the power developer were committed to hearing both sides (a decision that, in a way, diminishes the authority of both). The report continues: "Both have stated that if appropriate ceremonies were financed by AESNP and carried out by themselves, the spirits will accept project-induced changes to the spiritual landscape."[1]

Whatever exactly happened, we know that the two diviners summoned the spirits and asked them to relocate to a place that would be safe when these banks were inundated. Jaja says the spirits never agreed to move. Nfuudu says they did.

My visit to Uganda was nearly ten years after those spirit-relocation ceremonies, and, as I grew more and more curious about the story of the dam, I visited the spot where the relocations had taken place. I hired a motorbike taxi in Jinja and we traveled north out of town, on the highway running parallel to the river. After a while the driver turned onto a hilly dirt road flanked by the waxy bright green of banana trees. We zipped down the hill through clouds of red dust, the bike stuttering and slowing as we climbed up the other side. Eventually the road leveled off and the forest gave way to rows of small tin-roofed houses, the women inside them selling soda, bracelets, fruit, batteries, and phone cards, or roasting meat to serve with steaming portions of matoke, Uganda's ubiquitous plantain mash.

In addition to spiritual leaders and hydro developers, the visible power of that stretch of river had attracted the attention of whitewater rafters, who flocked to Jinja from all over the world to experience the rapids' energies. Beyond the vendors a steep staircase of stones and tree roots wound down to the riverbank.

I climbed down the steps, past the rafters' cabins, past the row of rustic bathroom stalls, past the open-air thatched bar where tanned Australians sipped Bell Lagers, to the launch area. A group was getting ready to start their trip down the river. Six rafters waited, tense in their helmets and life jackets, cushioned from the rocky shoreline by layers of inflated vinyl. Their guide waded in up to his shins and gave the raft a shove, hopping onto the rubber inflatable just as the river caught hold and whisked it downstream. In seconds, they were around the bend and out of sight.

It was easy to see what made this place so special, how it drew so many different types of people to it. I'd never seen so much water exerting so much force. The air was full of mist, and constantly moving. The ground itself seemed to vibrate the soles of my feet—maybe it did, or maybe it was just the effect of the roar of the water. But all these elements—the sound, the thrum of the bank, the cool wafts of fog—were peanuts compared to the river itself, an arc of pure force carving its way through stone. The Basoga, the whitewater rafters, the hydropower company were all responding to the same awesome, encompassing energy. Their responses to that power were different: to worship it, to ride it to an adrenaline high, or to wrest it into submission and transform it into the domesticated form of power that runs our lamps, the fridge.

I walked farther downstream, to where the riverbank curved in a protective S shape, creating a small inlet of calm. Two young men had docked their wooden rowboats with outboard motors, offering trips out to the islands in the middle of the river. I paid for my passage and hopped into a rickety boat along with two young Danish women. The driver barely used the motor at all during the ten-minute ride. Talkative on the shore, the boatman was quiet on the river. He focused on navigating, his relationship to the water at once cooperative and adversarial. He seemed to know the spots where the current was gentler or even shifted direction entirely. Using a long paddle, he eased the boat through this network of secret swirls and ebbs and flows so that we moved slowly toward the center of the river instead of being pulled straight downstream.

He pulled the boat up to the lee side of the small island, a mound completely covered in thin-limbed trees, which were in turn covered in ferns and climbing vines. We climbed over rocks and up a path to the top of the island, about twenty feet above the water. "This island," the young boatman shouted, "is a spiritual place, used for special spiritual purposes." His voice barely carried over the noise of the water crashing into the boulders around us on all sides. "Who uses it?" I shouted back. "What kind of rituals?"

"The spirit of Bujagali, he walks here, on the top of the water," I thought I heard him say, but I wasn't sure I had heard him right over the roar of the river. Basoga tradition has it that, when a new Nabamba Bujagali assumes the title, he does so by walking across the rapids—reading about it, I had supposed it was a metaphor. I didn't want to interrogate the young boatman with incredulous questions about religion. But I knew who I had to talk to.

Before leaving the river, I stopped again at the bank. It was hard to look at the current and not imagine oblivion—the edge of the water had the siren song of a cliff. Mesmerizing, deadly.

Yet in that huge, muscular cord of water, small birds were diving. Scrawny cormorants with long curved necks that gave them the silhouettes of snakes floated along in calmer portions, jauntily diving for fish. Earlier on my trip, some boys in town had tried to impress me, shown me a cormorant they'd captured and clumsily tied up with twine. It was wet and wriggling, on its side in a cardboard box, and I'd squealed with pity and said that they should let it go. Here were its cousins, casually diving through water that set my human heart trembling. I felt high on the river's power, felt I could see the power that was being exchanged and exerted everywhere, usually invisibly. Each bird that dove, I thought would certainly be smashed against an underwater rock and drowned—I laughed with relief and awe as each one surfaced and resumed its unbothered bobbing.

I arrived at Jaja's home one morning and was welcomed past a low, cracked concrete wall into a large plot of land with the brick frame of a large but incomplete house at its center. Some rooms were covered with sheets of corrugated metal, but others were open to the leafy tree branches above. A disused car sat in front of the house, several small goats picking at the grass around it. Beyond the goats was a series of squat earthen structures: shrines to house the local spirits. It was a humble place for the man on whom the last hope of the anti-dam movement was pinned.

In the yard, Jaja's son Moses greeted me and my translator, Anthony. Anthony and I had met and become friendly at the loud parties that were always being held in the backyard of the guesthouse—he'd grown up in the Basoga villages and knew the language and the religion. He usually led safari tours but agreed to be my translator and help me make appointments, even though he was always saying he was mystified by my interest—the traditions were old and boring to him. Moses led us inside to a living room stuffed with two massive armchairs, a sofa, and a coffee table. Jaja was visible in the other room, sitting on the floor with three women, all four of them shredding their way through a large pile of dried leaves that would be part of a medicine, Moses told me. Moses had short hair and wore Western-style slacks and a white button-down shirt; he would have looked at home walking along the modern streets of Jinja or Kampala, but his father, Jaja, had long, yellowing dreadlocks and wore over his shirt a tattered barkcloth vest covered with cowrie shells.

Jaja came out of the other room and produced a large reusable shopping bag emblazoned with a picture of a young couple kissing, which served as his filing cabinet. I asked him about what sorts of spirit-relocation ceremonies he had participated in over the years, but he wasn't interested in these questions and seemed intent only on providing me with evidence that he was in fact the real Nabamba. He pulled out a booklet produced by the World Bank Inspection Panel detailing its mission and featuring a photo of himself with some members of the panel. He showed me notes addressed to "Mr. Bujagali," inviting him to various ribbon-cutting and groundbreaking ceremonies that had taken place over the past twenty years.

I hoped he'd produce one particular 2001 letter I'd read a copy of in a World Bank report, sent from a public relations firm to request the initial appointment between the Nabamba Bujagali and representatives from AES Nile Power. The letter declared its purpose to "humbly request an introduction meeting between the Living Bujagali and Shandwick UK Public Relations Consultants."

I'd been fascinated by the collision of worlds that letter represented—the world of international power development, speaking a language of corporate public relations jargon, and a tradition with a relationship to nature in which the highest spiritual role was the guardian of a waterfall—and the same unexpected mixing was visible in Jaja's house. We sat on furniture that could have been pulled from the floor of a Raymour and Flanigan, looking through a gaping hole in an unfinished brick wall at a shaman shredding leaves for medicine.

This juxtaposition of tradition and newness also showed up in the way that people talked about the spiritual leaders of Bujagali. When I talked to a young reporter friend in Kampala about going to see Jaja, he had laughed and said, "That crazy one with the dreadlocks?" When I told one of the brothers running the guesthouse where I was staying in Jinja, his face turned suddenly very serious and concerned. His response made me nervous: he warned me to be very polite to the Nabamba to avoid incurring a curse. He advised me to bring some kind of gift, like maybe some eggs.

I did not bring eggs (Anthony had been mortified by the egg idea—he had a general horror of the rural—but hadn't suggested something else). I did make a donation to a basket of cash that one of Jaja's wives brought out. Jaja didn't want to talk about the ceremonies of the past. He preferred to discuss his plan for an ecumenical meeting. He had not been able to get the spirits to agree to leave the falls and was unsure how to proceed. This made sense to me, after visiting the river—the power was the place. I didn't know much about the spirits he was talking about, but if they had something to do with the awareness I'd felt at the riverbank—of my own smallness, of the way that force reverberates, of the secret quiet parts inside of tumult—I understood why they could not just be packed up and moved somewhere more convenient. But he was sure he could figure out what to do next; he just needed to bring together all the other major spiritual leaders from Uganda and discuss.

From the shopping bag, he pulled out a proposal requesting several hundred thousand dollars of funding for the conference, which would bring sixty-five spiritual leaders together. "My interest is not financial," he insisted—reflecting the fact that, as international controversy had grown around the dam, each side accused the other of bad faith and bribery. "It's just that I want this all to be done." The conference would help him and the other leaders figure out how to proceed, to appease or persuade the gods and avoid tragedy that would result from the dam going forward without divine approval. "People working on the dam would die," he said. "I'm very worried."

In fact, the entire delay was about money, and the steep tag for his proposed ceremony reflected his experience that the World Bank and the entities involved in the hydropower operation had not just deep but bottomless pockets. Indeed, it must have seemed that way: the price tag for the dam project recently topped $800 million U.S., an amount that expressed in Ugandan shillings is so high as to be essentially nonsense to someone who makes a living dispensing medicine and advice for donations nearly always under a dollar. Since the 1990s, Jaja's neighbors had been offered cash and plots of land for their own resettlement; he was given money to perform ceremonies to relocate spirits that he didn't control. The way that developers deal with cultural property and spirituality had become clear: monetize it.

I had not, until that day, stopped to think about why or how religion might be one of the last possible ways to stop a dam from displacing the Basoga. Why did this extremely old man, a frail figure in a falling-down house outside of town, seem to be the last bastion against a collection of international corporations and financiers? The fact is that, in guidelines governing World Bank financing rules for such projects, designating a property as spiritual or cultural is one of the few ways that spaces can be safely removed from the realm of what's for sale. There are no special provisions for land or water that is especially beautiful or biodiverse. You cannot block a giant infrastructure project from inundating your home by simply saying that your family has lived there for generations and that you prefer to stay.

A few weeks earlier, I had visited the offices of the existing dam at Owen Falls. There, I was asking the engineers and operators about another pressing question of mine, one that had nothing to do with Bujagali Falls. I wanted to know more about the missing water in the lake. Lake levels had started to recover, since their lowest point a few years earlier. But it was still unclear why they had dropped in the first place. A hydrologist I'd met in Nairobi told me he had been modeling the lake level for years. He had factored in municipal uses, rainfall, drought, temperature—everything—and had concluded that the missing water could only be explained by operators of the Owen Falls dam releasing more water than usual to generate more power.

If true, his conclusion would have international implications, because water-release protocols were bound by an agreement among nations that share Lake Victoria (Kenya, Uganda, and Tanzania) and many more that share the Nile, which the lake feeds. The dam operators insisted that the amount of water flowing through the dam had never exceeded the levels legally determined by "the agreed curve." But, in between meetings, they also left me alone in an office with a poster, possibly not realizing that it contradicted what they had just told me.

This wasn't the first time, of course, that struggles over electrical and political power intersected. When the Owen Falls Dam was completed in 1954, the United States and Britain had offered to fund the construction of the Aswan Dam, on the Egyptian Nile. But in July 1955, Egypt's President Nasser arranged an arms purchase from Czechoslovakia, ending Egypt's dependence on Western weapons and angering the British, who saw it as an overture to the Soviet Union and withdrew their offer of financing. To pay for his dam, Nasser nationalized the Suez Canal, the passage between the Mediterranean Sea and Indian Ocean that had since 1904 been a neutral zone. Nasser took control of the canal and planned to use toll revenue to complete the dam at Aswan. In October, France, the United Kingdom, and Israel invaded Egypt.

During this time, British military officials stationed in Uganda developed a plan to use Owen Falls as a piece of weaponry. They calculated that by decreasing the outflow through Owen Falls by about 85 percent they could in effect shut off the Nile, at least during the winter and spring, when the Blue Nile would run low. Memos were passed back and forth between intelligence agencies and the prime minister, who was planning the shutdown. I kept thinking about this plan, about the ways that geography and hydrology make politics, about the many ways that power can be harnessed or transformed. Ultimately, the plan was abandoned, as it would have been too slow-moving as an act of war: it would have taken sixteen months for the plan to affect the lower reaches of the Nile. It also would have put much of British East Africa underwater.

A few days after meeting Jaja, I visited his rival, Nfuudu. His compound was farther from the river, on a road that leads out of town to the nearby army barracks. A tidy hand-painted sign gave his full name and title: Mzee Nfuudu Benedicto Badra Jjaja wa Lubaale Kimaka ("Sir Nfuudu Benedictor Badra, Grandfather of the Spirits of the Lake"), and listed the price for a consultation as two thousand Ugandan shillings, about ninety U.S. cents. The yard was covered with a layer of leaves drying in the sun—leaves from the same medicinal plant Jaja had been shredding. When I arrived in the morning there were already six men and women sitting on folding chairs on a cement platform in between the long rectangular structure where Nfuudu and his family lived and the small earthen dome-shaped shrine where he met with clients. He was too busy to see me that day.

I returned a few days later. Nfuudu met me wearing the same pressed pink shirt and trousers he'd been wearing when I last visited, but before going into the shrine, he changed into a long barkcloth vest and wrap covered with shells. The difference between the two men's financial situations was immediately clear: like his compound, Nfuudu's ceremonial clothes were in better shape than Jaja's. Inside, the shrine was dim. Our interview followed a similar pattern: he mostly wanted to impress upon me that he was the real spiritual leader. When I asked him about the older Nabamba, he laughed and told me that "the spirits cannot work with dirty people." He described moving Lubaale, one of the highest spirits, from the falls to this compound. "There are fifteen shrines here," he explained, "and now Lubaale lives inside of them." He made the same accusation that Jaja had made of him: that the other Nabamba was a false leader, that he was just telling lies so that he could make money.

I told him how Jaja had described his relationship to the spirits differently, saying that he was only a medium and that the spirits made up their own minds about when and whether they would move. He laughed and said, "Of course I can move them. I am like the president to these spirits," and banged a long beaded pole on the ground for emphasis.

At one point, while he told a story about the 1953 ceremony that opened the Owen Falls Dam, I asked him to slow down to make sure my translator could get everything he was saying. "I want to make sure I have direct quotations," I said. He stopped talking and looked at me for a moment, nodding. Then he rolled his eyes back in his head and shook his head and torso, and let out several high-pitched whoops. He looked at me and delivered, in a deep rumbling voice, a few lines my translator translated as "It is me. I am here. I am the one, Lubaale, the spirit from the falls." Nfuudu, it seemed, was momentarily possessed, or at least acting that way—perhaps misunderstanding that I wanted direct quotes from him, not from the spirit realm.

This seemed to perfectly exemplify the difference between

the two Nabambas: Nfuudu would enact spiritual possessions, and Jaja would not. Nfuudu said he had power over the spirits; Jaja said the spirits had power over him. With the controversy around the dam, both had found themselves in a position to liaise with wealthy strangers who wanted something from them. One was able and prepared to give that thing, the other not. The veracity of either man's claim was still unclear to me, but it started to seem almost beside the point. As a practical issue, it mattered little who was right. The dynamic between the two of them, and between them and the hydro developers, demonstrated a larger phenomenon happening all around the developing world—at all of the front lines of extractive industry, really. Things were changing: those who could find a niche for themselves in the giant, internationally financed systems would prosper. Those who can't—or choose not to—won't.

The Bujagali Dam was completed in 2011 and started operating in 2012. Three thousand households were relocated or lost property when the reservoir filled in over their land. In hindsight, it seems inevitable—the taking of indigenous land for such projects is a worldwide phenomenon, one that continues even if it sparks outrage. Even when I was there, when the controversy was swirling, with international and local environmentalists petitioning the Ugandan government and the World Bank, the machines were already at work, plowing ahead in spite of the confusion and disagreement. I think often about one of the American executives at the hydropower company, whom I'd interviewed in his office at the construction site. I'd been asking him about the two spiritual leaders—these two rivals were, in the view of the developers, simply a hurdle on the road of modernization. Their competing claims to the spiritual title were just so much backwater nonsense. He had laughed loudly, leaned all the way back in his chair—so far I thought he'd top over backwards—and said, "You want to write nonfiction, and you came here. What a fairy tale."

That comment haunts me. It embodies an idea that's truly sinister—that the worldview you don't understand must be nonsense. That if something's worth is established according to a calculus you don't understand, it must have no value at all. This is how cultural misunderstanding gets spun into exploitation, erasure, theft. I didn't, during my months in Uganda, ever fully come to grasp what the Bujagali Falls meant to the Basoga. But I do think I glimpsed it a few times, imperfectly. It isn't mine, but I certainly saw and heard enough to know it was real.

Jaja Nabamba Bujagali died in October 2019 and will be remembered as "the oracle of Bujagali." He was, according to the news reports, 113.

Notes

1. The Inspection Panel, Report and Recommendation on Request for Inspection: Uganda: Third Power Project (Credit 2268-UG) and the Proposed Bujagali Hydropower Project, October 10, 2001, www.inspectionpanel.org/sites/www.inspectionpanel.org/files/ip/PanelCases/24-Eligibility Report (English).pdf.

MOBBING CALL

TRACY O'NEILL

When we were still in a season of firsts, the last man I brought home to New Hampshire and I were people who stayed up too late talking in awful railroad apartments and after-hours spots that would later become, in memory, always dark yet glowing blue. As far as he could have known, I only ever wore sequined cocktail dresses and elevator shoes. I always saw him in Ben Sherman suits. It wouldn't have occurred to me to ask what sort of family he imagined for himself. For the most part, the furthest into the future I spoke were vows to vacate my lousy nightclub job. It took me falling asleep at a party in the middle of a conversation, then waking to him removing a cup from my hand as he covered me with a woven floor mat, for me to see that he wanted to take care of someone.

Maybe it was that improvised tucking in, or maybe it was the crushing light in his face as he recalled after-school snacks of pickle sandwiches that made me want to show him where I grew up. We got the honking silver pickup on the highway. I got grandiose and I meant it. I told him the New Hampshire of my childhood was one in which the natural world supplied a sensory rampancy approaching an ethic of liberty, that there you are free to perceive with dilettantish abandon. I told him that where I grew up there is Baboosic Lake shifting blue at its center through the summer, then the red smack of turning leaves. In winter, you feel the cold, sickening fist of glee bobbing up the gullet in a slide over ice—and, if you are lucky, someone's mother pouring hot maple syrup into packed snow to make candy.

I told him that to be precise about New Hampshire requires paradox. It can be summarized by its poles. The state motto is "Live free or die," and therein lies its conceptual sexiness, its blunt-force frisson. It's in this grandiosity you can cultivate ambitions to erect a cathedral to zero fucks or die trying, even as it inspires an obsession with dwindling time, too. New Hampshire's spectacular parade of seasons may mean you reside in pointed existentia, suspect it is more accurate to say live free *then* die; but by all means, live free first. How many more autumns will I see the flame of leaves? you may wonder, I told him. When, if it does, does the world end?

If it is not evident, this man was patient with me.

On that first trip home to New Hampshire, it was presumed that this man was going to be the father of my children. He brought a basket of Hudson Valley jams to my family. We hacked crusted ice from the walkway to the door. My parents noted that we both had thick, dark hair; were good eaters; additionally, he was broad where I was slight. And despite an ambivalence toward parenthood, it had become somehow a given that if we were to have a child together we would name the child Sammy. I was stupid with love, and in those moments, I could believe that to stand proximal to childhood was to be availed of astonishments, to feel the bright firstness of a child's experience smear back into one's own. I could believe in the green of Baboosic Lake shifting blue. The father would be the one to teach Sammy to swim since I was hapless in water.

Yet even as we visited New Hampshire, even as my friends celebrated the third birthdays of their children and so on, there were doomy tolls sounding. One day, I saw a news item on the internet saying Nashua, New Hampshire, had been named one of the safest cities in the United States to raise a child. Nashua abuts the town of Amherst, where I grew up. Accompanying the ranking were tips on how to sustain the safety of one's child: Practice crossing the street safely. Monitor their online activity. But what struck me most was Tip #8: "Establish and practice a family escape plan." How, I wondered, would a family escape climate change?

It is difficult to fix the discovery of climate change to any one date. It's been studied in some form since the nineteenth century. Most people recognize the 1980s as the decade in which the urgency of human-caused global warming solidified into scientific consensus. And as early as 2003, even the Department of Defense acknowledged the threat it posed, publishing a report referring to it as a "plausible" if "unthinkable" security crisis. In fact, it was not unthinkable. The Pentagon thought quite seriously about this potentially operatic tragedy: how commercial fishermen would be insufficiently prepared for the migration of their quarry, how some pest populations would thicken through the altered weather, how food scarcity would disrupt economies, and how, eventually, conflicts over dwindling resources might drive interstate violence. A non-negotiable number has been named. The UN estimates we have only until 2030 to forestall disaster.

The notion of the ticking clock is not unfamiliar in narratives of reproductive adversity, which remind prospective mothers at dizzying turns that even the happiest childlessness will not always be only a choice, will one day become an unavoidable fact. We accept the science that tells us fertility is finite and revocable. But many of us today have honed a selective acceptance of science, believing instead that when it comes to climate change, there will always be another Tuesday. It is a popular delusion that, under liberalism, deferring action to avert environmental catastrophe is an exercise of freedom.

My personal deferrals have tended more toward the child question. In the years that followed that trip home to meet the parents, I spent days at a time reading in Brooklyn. I said, more often than not, yes to one more party before going home. Apropos of a cigarette, women would tell me they quit *like that* the moment they got pregnant. My mother began referring to my pet as her grand-dog. I enjoyed the kind of attentive friendships in which we did not merely ask after the kids, asked at least as much, "How are *you*?" All the while, New Hampshire was, of course, rich with admonition of time passing. Snow fell. Snow melted. Ice hardened on branches like glass gloves, and then came new shoots. When the last man I brought home to New Hampshire expressed enthusiasm over some kempt, expensive place, it might have been, confoundingly, because of the school district.

Confounding because I had, indeed, fallen into imagining the life of an unborn child. That speculative biography consisted of the child at twelve or eight or some other number knowing not only that their personal experience of the world, but the world as we know it, was numbered. Maybe there would be the warring consequences predicted by the Pentagon. Definitely there would be flooding and heat clumped with disease-carrying bugs. The child would become someone who was twenty and afraid, thirty and afraid, fifty and afraid. I would have brought into being a consciousness who would experience terror for which I could offer no consolation. I was not sure I could abide this story, let alone cause it.

The essence of a story is not constituted by its worst events, but I wondered what of the landscape would remain to inflect a life with pleasure. The New Hampshire Department of Environmental Services has concluded that autumns will go duller in color, green ceding quickly to brown as trees die off, non-native species invade, and flora becomes "climate stressed." As temperatures rise, skiing season will be truncated. Others anticipate that the warmed weather will rob the maple syrup yield. The possibilities for licentious New Hampshire experiences will narrow, becoming less available or not available at all.

Happiness hardly requires leaf peeping and winter sport, of course. Plenty of people manage all right without them. But even those parents I know who do not particularly love parenting want excess for their children. They want the better school, the better food. Even before the children were born, they wanted the best ob-gyn. They are prone to declaring they want to give their children the world. If this is true, then the question is why we've not collectively wanted to preserve more of what makes the world worth living through.

What if we cared not by producing new sentience, I asked the last man I brought home to New Hampshire, but by sustaining our world for those who already and may one day exist? Robert Frost put it differently in his poetry collection *New Hampshire*. Considering what sort of life to fashion, the poem's speaker declares, "I'd hate to be a runaway from nature."

In 2017, a Lund University study found that the most environmentally destructive action an individual can do is to bear offspring, a choice that would require 684 teenagers to commit to comprehensive, lifelong recycling to offset it. I read this study when I was one year away from being someone whose pregnancy would be, in medical parlance, "geriatric." From friends, I solicited counsel on what it meant to reproduce into likely devastation. Almost invariably, someone would say, "But that's politics. What do you *want*?" I did not understand how desire would exempt my hypothetical child from climate change.

The next year, a friend gave me a small novelty book where I encountered the notion that New Hampshire has an "ungenerous," meaning infertile, nature. It is a rocky place, tight in its crop yields, and in the book, it was said that the White Mountains, in their rigid looming, "signal something a bit aloof about the state's character—a rectitude that stands in splendid isolation." But to me, New Hampshire has never been aloof. Its rectitude, if that's what it is, simply remains agnostic. Though enamored of freedom, the state remains undecided about how exactly the human prerogative fits into it.

The regional liberation narrative is, after all, defined by a particular strain of one-or-the-other. New Hampshire was the first American colony to assert independence from Great Britain in January 1776, unless you believe that it was Rhode Island in May of that year. It wasn't much of a battleground in the American Revolution, unless you believe the work of Portsmouth pirates counts. Exeter, New Hampshire, is where the Republican Party was established on October 12, 1853, unless you believe the party was born in Ripon, Wisconsin, on March 20, 1854. And after his son's death under the watch of a white family, Chief Chocorua cursed the settlers in revenge before dying of gunshot wounds, unless you believe he cursed them and then threw himself from the peak of a mountain to his death. From a view on high, it might appear that we are not especially adept at determining the arc of where living free or dying begins and ends.

My friend gifted me the book shortly after it became apparent that the last man I brought home to New Hampshire and I would not have a baby together. In several ways, it seemed, it was too late, and in turns of ironic thought I remembered how my mother used to fear I'd become a mother too early. Hers was a family where habitually unwed women became unwed mothers. That fact had not convinced her that life went on but that freedom was easy to ruin.

I recall debating such potential disastrousness with her on the telephone one afternoon when I was younger. That day, I pictured her by the glass sliding door of the family living room. She would have been turned to the window, a cordless telephone clamped between her head and shoulder, looking out into the thick pines behind the house, the ones that went black at a particular deepening hour when I was small and terrified and thrilled, in a race against an uncanny sense I was being pursued.

"But so what if life is beautiful?" I said to my mother.

My point lay somewhere in the day when I was a child and it blizzarded hard enough that the weather came up to my chest. I'd had to swim my arms out first to dig a path. In one moment, my mother let out an awkward, ugly scream, believing I'd drowned in the drift, though I was only lying in a self-made snow grave. I was only trying to catch sight of the six points of snowflakes before they expired on my face. I could not quite make out where one point became another, or if the snow was soft or stinging as it came down on me. I stayed there open to the sky, half-forgetful that I'd intended to impress angels in the yard, happy.

"Isn't new life always beautiful?" I said to my mother, because I wished it to be true. I still do.

MOMENTS OF BEING

KIM STANLEY ROBINSON

When the editors asked me if I had ever written anything about personally witnessing the effects of climate change, I said no. I was thinking that because weather has always been wildly variable, climate change would always remain at the level of the statistical, a scientific finding that's hard for some individuals to see.

But then I realized that wasn't quite true, in my case. I've been going to the Sierra Nevada of California for almost fifty years now, and high mountains, like the polar regions, are warming up and manifesting impacts of climate change faster than other regions of the Earth, for reasons not fully understood.

So, two things: The Sierra Nevada has about one hundred small glaciers and four hundred "glacierets," glaciers so small that their ice doesn't slide downhill. Even the biggest remaining glacier in the Sierra, the Palisade Glacier, is only about a mile long, and it's lost about half of its ice mass since 1935. That ice loss is like a thermometer, in that the rise in temperature causes something you can see with your own eyes; and I have. Photos I've taken of Sierra glaciers over thirty years show very clearly how much ice they've lost. Soon all the ice in the Sierra will be gone. This loss will be bad for some meadows, and for the look of the range, as far as humans are concerned; in the Himalaya, glacier melt will be catastrophic for almost a billion people, because glaciers provide their primary water source.

That's one thing. The other is this: in the last several years, it seems as if the summer monsoon, which used to come out of the Gulf of California and rain on southern Arizona over a few weeks' time, has been extending northward and striking the Sierra Nevada more often than in years past. Articles in the scientific literature confirm that this is the case, and not just our unhappy anecdotal impression.

That makes backpacking in the Sierra pretty dramatic. Summer weather in these mountains used to be famously benign. Among members of the mid-twentieth-century Sierra Club, it was a truism that you didn't even need a tent in the summer Sierra, because it never rained then. People laughed at this, knowing by experience that it wasn't totally true. But it was a tendency. California has a Mediterranean climate, and lower parts of the state rarely see rain from late April to late October. In the high Sierra in those dry months, thunderstorms frequently cook up on hot days, soaring from small origins just over the peaks to heights of thirty or forty thousand feet, where they become billowing masses. Then comes a dramatic hour of black cloud, lightning and thunder, and a downpour of cold rain, or even hail, after which the sky quickly clears to a glorious sunset.

That was normal. The appearance of an Arizona monsoon was not. Only once or twice in my life did the Sierra in summertime get visited by humidity, thick low clouds headed north, and irregular but copious all-day rain. In this last decade, however, it's been happening more and more often.

In 2013, Carter and Darryl and I were at the elbow of Seven Gables Canyon, setting camp in a cloudy sunset. Earlier in the day we had stood under trees waiting out a downpour so intense that the ground of the entire canyon suddenly turned to water under us, as accumulated rain poured down the tall gleaming side walls on either side of us. We had to clamber awkwardly onto big tree roots to get above this flood, which though only a few inches deep covered the whole canyon floor. We'd never seen anything like it. After a while the water drained into the roaring creek and we continued to pick our way up the canyon.

Now we stood in an open sandy meadow, very wet but nice for camping. We set up our three tarps and bedded down in the cloudy dusk. That night it began to rain again, then to blow hard. The wind seemed to be coming from the south, down the canyon, until it whipped through the elbow bend we were in and headed west. It threw rain against our tents in what sounded like sprays from a fire hose.

There was no sleeping that night. Our tarps were barely holding on. Mine seemed to have water running up its inner walls, perhaps under the pressure of the wind. And I was camped on thin grass, growing out of the usual Sierra decomposed granite, which this time was maybe contained in a shallow bowl of bedrock, a pretty common situation; now rain was coming down so hard that my bedrock bowl was filling with water above its level of sand. I was beginning to flood from below.

My air mattress floated me above the shallow pool of water growing under me. This is another great thing about modern ground pads: they float, and they're pretty tall. So this was the least of my problems. Rain was still bombing down, wind hitting in violent slaps. Something could very possibly give—a tent stake in the ground, a loop holding tent stake to tarp—even the nylon fabric itself could split. If the tarp failed, my widespread incidental dampness would change to total fundamental wetness instantaneously. I had to plan for that, and I did; it passed the time, sleep being out of the question.

When the tarp gave way, I decided, I would quickly put my rain jacket and wind pants on over my night warmies, get into my boots, wrap my down bag and failed tarp around me like a burrito, and sit with my back against the lee side of a big tree, waiting for dawn and the end of this incredible downpour. It could be done, if I had to do it; it would not be the worst thing that had ever happened to me in the mountains; etc. In my mind I was prepared to march through all the necessary steps. Meanwhile, the storm lashed us.

In the event, all our tarps held. The wind relented, then the rain stopped. When the sleepless gray dawn came, we crawled out completely bedraggled. We consulted with each other. We'd all had the same experience, so we didn't have to describe the particulars, merely marvel at them, pretty pleased we had done as well as we had. We concluded we could maybe use stronger tarps. Carter had heard of a new fabric called Cuben Fiber, said to be more waterproof than nylon.

•

The next year we had new tarps made of this fabric, now called Dyneema. On our third day we climbed to the pass between Sixty Lakes Basin and Gardiner Basin. Mid-afternoon on a warm muggy day, with some clouds overhead. We had been rained on during both the previous days, so we suspected the monsoon. Even so, we were surprised when in about fifteen minutes the sky went from white to dark gray, and suddenly it began to rain, then hail, then thunder loudly. We didn't see any lightning bolts, but the clouds over us turned white right before each big rolling boom. Time to leave.

So we put on our rain gear and started down. The descent into Gardiner Basin is steep and rocky at first, a wall of big boulders that one has to traverse, down toward the highest Gardiner lake, a long skinny thing extending as far as we could see, especially in the rain; the rest of the basin was out of sight below this endless lake. The rain was hard and cold. The clouds lit up over us, then immediately banged. Their bottoms were scraping over the ridge to our right, and the boulder slope we were descending was slippery. It was like an immense, long room: two gray rock walls, between them a long gray floor of lake, under a pelting rain coming out of a gray ceiling of sky. We were inching along the bottom of the right wall, where a narrow band canted into the floor.

I was using one of my blue backpacks, the ones you could cut with your fingernail. I love them, but they're not waterproof, and my gear inside this one was stuffed into stuff sacks that also were not waterproof. We had our new tarps, and it might have been possible to pull them out and figure out a way to use them as capes, or at least as backpack covers, but with the thunder now cracking overhead, and no shelter anywhere, and the rocks so very slippery, and everything so cold and wet, we didn't feel like stopping to make experiments in gear deployment. I was going to have to deal with what I found in my pack when we found a place to camp. Even my wool mittens, which would have been great to have on in the cold of this storm, were packed too far down in my backpack for me to want to stop and get them out. I pulled my rain jacket down over my hands as far as I could and grasped the sleeves against my walking poles, leaving only my fingers exposed. Having cold fingers was not a major problem. As always, I wondered if aluminum walking poles were in effect little lightning rods. It seemed like they could be. It even seemed like they were humming, although I confess this sound could have been just in my head, an electric buzz like wet tinnitus. Probably the keening of the wind.

The descent needed care, and I gave it. The ridge above us to the right was now completely lost in the cloud. The thunder was getting less frequent, the rain harder. The highest lake in the Gardiner Basin was turning out to be really long. Before the storm struck, we had been planning on hiking to the next lake below it. Now I was thinking that maybe the flat spot I could see at the far end of the lake, where a little promontory pinched off the endless stretch of water, might do. It would be very exposed out there, the cat's paws flying across the lake's beaten surface made that clear, but it looked flat, and nothing around us was flat or even close to it; it was a slope of rocks between the size of refrigerators and bowling balls. Nothing to do but clamber down and across this jumble, getting wetter and colder.

This was one of those hours where you just have to bite it

and forge on, coldly determined. Total exposure to the elements. Lear on the heath; Beethoven's mad blind energy, as in the *Grosse Fuge*, or the end of the *Hammerklavier*. I think Beethoven must have gotten caught out in storms once or twice. Often I like those hours, even while they're happening. It's like sticking your finger in a wall socket, but at a level of electrocution that you can stand. Hammered by the elements—unsheltered—focused—at one with the world. Forge on!

Then, as I was getting closer to the little ridge that pinched off the end of the long lake, there appeared at the start of the pinch a little triangular patch of grass. Real grass, and almost flat—certainly not the biggest tilt we had ever camped on. And it seemed like it might be just big enough to fit our three tents side by side. This was like one of Piaget's tests of cognitive development for toddlers: Was the patch big enough for three tents? Maybe?

When Carter reached me, I proposed the plan. He nodded. Good idea, he said. I think we'll fit. And if we don't, we still have to.

Darryl joined us and we sketched the idea to him. He nodded. Let's do it, he said. We were all drenched to one extent or another.

In the pouring rain we set up our tents right next to each other. You could barely walk between them, as our tie lines overlapped each other. I was in the lowest part of the patch, and set my zipper door (so fancy, a tarp with a real zipper!) on the side away from Carter.

I got under my new tarp, sat down cross-legged under my poles, and began digging in my backpack to assess the damage.

As I pulled things out, I grew more and more appalled: everything was wet. Incidental dampness was long gone, although there was a lot of that too, but mainly it was a serious case of fundamental wetness. All my socks were too wet to wear in my sleeping bag. My blue warmies were wet. Down vest was only incidentally damp. And the sleeping bag was only wet in places, but where it was wet, it was soaked.

It was only about three in the afternoon, which gave me time to take full stock of the situation. My new Dyneema tarp was pale green and translucent; I could see the rain running down the outside of it, but as I touched the inside of it, I could tell how dry it was. And it was big—well, bigger than what I was used to. It's not actually that big. But the light poured through it. And I wasn't in the rain, which continued to pound down. The grass under me was wet, but that was incidental dampness; with my ground pad inflated I was well above that, and my gear was resting on my rain pants or jacket, or something else waterproof or irrelevant to my night life.

So I lay on top of my sleeping bag, which was draped on the ground pad, and figured I would sleep barefoot, since I had to. The tilt was pretty steep for a bed, but it was cleanly tilted head to foot, not side to side—I had set the tarp to get that. And the exertions of the previous hour or two—I found I had no idea how long our descent had taken, but probably it was more than an hour, and maybe less than two—that work had warmed me up, all but my fingers. I could eat, drink (the long lake was not far from our patch of grass, and we had grabbed water for the night), and rest. Listen to the rain drum on my taut new tarp, and even watch it running down the other side of an impermeable translucent barrier, in the usual infinity of delta patterns. I was sheltered! Over the next couple of hours I began to settle in to a deep sense of safety, comfort, and even warmth. Damp warmth, but so what! Warmth is warmth, and I wasn't going to get any wetter than I already was. This new tarp was not just a refuge; it was a miracle. In that hour I fell in love with that tarp. I was home.

The following year, we went up Center Basin to see if we could follow the old route of the Muir Trail up to Junction Pass. It was easy; the trail is still very distinct, even though it has been abandoned and was taken off the maps in 1934. The old wooden sign in the pass, just a mile or so west of Forester Pass, is very evocative of the 1920s.

On our way back we decided to try to go over University Shoulder. This is just what it sounds like, a high traverse over the west shoulder of University Peak. We had seen the north side of this shoulder on our way in, and although it was obvious it would have been really tough to go up it, going down it had looked feasible. So I suggested to the others we try it—yes, it was my idea, and yes, I had failed to take in the sentence in the Secor guide that called this shoulder a ski route. Although, parenthetically, I am amazed at what backcountry skiers will think to try.

So we left the froggy pond at the bottom of Center Basin and took off up the slope to the shoulder. This slope is the side of a very big glacial canyon, the upper Bubbs Creek canyon, with the Muir Trail at its bottom heading up to Forester Pass. Soon we were using our hands to pull ourselves up, grabbing on to the exposed roots of small trees at head height above us. It was a vertical forest in decomposed granite, nobbled by big boulders that blocked our way but also stabilized the sand and gravel and trees between them. Ridiculously steep, and we could see, by looking across the big canyon to its other wall, just how slow our upward progress was. Most of two thousand vertical feet, at the same steep pitch. Frequent breaks. Already far more effortful than staying on the trail would have been.

Finally we got onto the shoulder proper, which gave us an easy sandy traverse across a high space where nothing but the big triangular pyramid of University Peak was higher than us, blocking our view to our right. In every other direction we could see forever. Joe and Helen Gompertz LeConte, and many other early Sierra Clubbers, had ascended University Peak from this shoulder in the 1890s, the women wearing long skirts, etc.; those people were tough. We were happy with our high traverse. Our canyon wall climb was almost feeling worth it.

At the far side of the shoulder, we came to a drop-off and could see down the steep slope we had declined to ascend three days before. A steep descent indeed, sand and scree dropping down a broad shallow funnel that at its bottom shot through a gap between boulders—that part looked a little nasty—but we could do it. So we sat on this fine overlook and ate our lunch and enjoyed the view. Which clouded up as we sat there. It had been muggy before, but the clouds had been mostly in the north. Now suddenly they were everywhere; they darkened fast; thunder rumbled.

Kevin Kline, playing the Pirate King in the film version of *The Pirates of Penzance*, said it best: Here we go AGAIN! Yes, I shouted his slogan as we poked into our rain gear. We hurried to get off the ridge and as far down the slope as we could before lightning started. It began to rain, cold and hard. Then lightning, thunder, hail, the usual dreadful combo; although again, the lightning was of the flashbulb-in-the-clouds variety, not visible bolts hitting the ground, which is the real heartstopper when you see it, those little elbowed lances of fire torching reality itself. This was just the usual awful booming overhead, with the whole world flashing in momentary flashbulb style before the gloom returned.

The steepness of the scree slope wasn't helped by being soaking wet, except maybe it was, as we could stomp into it with our boots a little. I sat down for a lot of this drop, in my usual style. Carter and Darryl, both good skiers, kept on their feet and stepped down skillfully, me bumping along behind.

We came to the boulders at the bottom of the funnel. I went down to have a look and discovered a drop of about fifteen feet, maybe a bit more; not huge, but for me an impossibility. It was pouring now, and the thunder frequently punctuated my cries of dismay. I shouted up to the guys concerning what I had found. Darryl immediately cut across the slope of the funnel to the boulders framing it on the left, to see if there was a way down through those rocks. Carter declared he was not to be stopped by any mere twelve-foot cliff, and came down past me to investigate for himself. I wished him luck on my way back up, then crossed the funnel as low as I could without slipping down its spillway and over the cliff. Darryl had meanwhile found a nifty staircase down the boulder wall to the left, and was already on the scree slope below the crux. Gratefully I followed him. Pausing on this boulder stairway, I looked across and was horrified to see Carter wedged in a crack partway down the little cliff on his side. He was in a nook that looked to be nine or ten feet above the steep scree under him—no way down, not easy getting back up, rain pounding us. I took a photo of him, hoping it would not be needed as evidence in any subsequent inquiry. He took off his backpack and tossed it down to the slope below. Holy moly! I shouted. Carter! The wind and rain and thunder were too loud for him to hear anything I shouted. Then, even though I was looking right at him, he was standing successfully on the slope below. He had jumped. I shouted again, amazed; somehow I had not seen his leap. But all was well. He put on his pack and proceeded. I did too. Darryl was far below, and the slope between him and us was straightforward scree, not as steep as in the funnel. We were past the crux.

But not. Secor mentions some "giant boulders" in his description, and I think these might be completely covered by thick snow in winter, so that as part of a ski route, they wouldn't be an issue. For us they were. These boulders were indeed giant, the size of garden sheds and school buses, all strewn in a crazy broad band that was no doubt once the lateral moraine of a really big glacier. The boulders had been there forever, and were covered with a kind of black lichen that gets extremely slippery when wet. Like thousands of flat lobes of slick black plastic. On top of these boulders, the gaps between them dropped into cellar holes sometimes twelve or twenty feet down. But if we tried to stay down at that basement level, the corridors would close off in little dead-ends with vertical sides more than head-high. Nope; it was a fucking nightmare.

The hail had turned to rain and relented a bit; the thunder had stopped. It was now just a matter of being soaked and cold and hiking over a wet boulder hell. Even though we were moving horizontally, and were no more than half a mile from the uppermost Kearsarge lake, where we had camped on our first night and knew exactly where we would set our tents, etc., we were now engaged in the most meticulously slow and dangerous work of the day. Though flat, it could still be rated class 3, in that a mistake could kill you; so we had to make sure not to make a mistake. From time to time this involved crawling. Many moves had to be made as slowly as we could make them. The wet black lichen was outrageously slippery. It was a bit like the *Batman* TV show of the sixties, where Batman and Robin pretend to climb buildings even though you can see they are on a flat surface and the camera is tilted sideways. A class 3 horizontal surface! Who knew?

The boulder field ran right into the eastern end of our lake, and our campsite, the only campsite on the lake, was at its west end. When we finally got over these last and worst boulders onto ordinary ground, for a last walk around the shore of the lake, we were soaked and beaten, but also relieved. Ordinary walking never felt so good. We were going to make it. When we staggered into our first night's campsite, it was like our long-lost home. We tallied up the day as we walked the final stretch; it had taken us eight hours to go two miles.

Recommendation: take the trail.

Not that I mean to generalize this! And it has to be added, Helen Gompertz in her trip report in the *Bulletin* of 1896 makes it clear that her Sierra Club group went up and down that very same slope when they climbed University Peak. They went over that boulder field twice in one day. And the route is listed in Secor's guidebook, so it is a real pass. But still. A nightmare. Well, it was because of the untimely downpour that it went a little haywire. That was what made it a day to remember.

May we have more days like it. And given this seeming increase in the frequency and severity of the summer monsoon, we probably will. Is this stronger monsoon an effect of climate change? Possibly so. Normally, that is to say, before about 2015, it was a humid July flow up from the Gulf of California, which for a few weeks gave southern Arizona a big shot of moisture by way of spectacular daily afternoon thundershowers. It didn't used to flow as far north as the Sierra, but now it seems that it does. There's speculation that it's an early effect of climate change.

Well, it could be worse. In fact, it will be worse. Drought is worse, and drought is coming; which means that these new extra summer rains in the Sierra might even be considered a good thing for the living biomes under them. In any case, we deal.

UNTIL THIS SNOW REACHES THE OCEAN

NICKOLAS BUTLER

High on a ridge at the northernmost extent of the Driftless Area lies our land. I think of it as a refuge—not just for wildlife, but also for our family. Our children, a seven-year-old girl and a ten-year-old boy, are constantly defining their world, its borders and edges; how far they can push and explore. There is a specific white pine they call *their* "climbing tree." They collect worms and bugs to feed our chickens. They hide on ridgetops above nearby roads to "spy" on passing motorists.

The boundaries of our everyday lives are well defined: these sixteen acres of recovering prairie and oak savannah are lined by barbed wire—our neighbor raises buffalo on hundreds of acres to the east and south of us—and asphalt, where two rolling country roads demarcate our property to the north and west. As someone with a dim vision of our future as a species, it is important for me to live in a place where my garden is expansive, where I can raise chickens, and, every year, where I can plant more and more apple trees.

But living here is also difficult. The year 2019 was the snowiest on record for Eau Claire, Wisconsin, with more than 101.5 inches of precipitation. It felt like the world was caving in. All around us, barns collapsed, killing hundreds, maybe thousands of animals, mostly cows and pigs. On many days there is the palpable sense of being buried, a feeling only heightened by the diminished light of this latitude. The winter of 2013–14 was the coldest on record, with an average temperature of just over seven degrees Fahrenheit. For weeks, the morning temperatures would be below zero. Allowing five or ten minutes for our cars to warm up became part of our daily routine. As I write this in the deep winter of 2019, the sun will set at 4:24 in the afternoon. Then, seventeen hours of total darkness. I recently learned that, if you look at a map of global temperature rise, one of the few places on our planet that you'll see actually *cooling* is the Great Lakes Region.

During these winter months, one coping mechanism—for the fortunate, at least—is to skip town and migrate south. My family and I routinely point our Subaru toward Hutchinson Island, Florida, where we spend about ten days every March thawing out, with the hope that spring may have arrived in Wisconsin by the time we return.

Hutchinson Island is a narrow barrier that lies to the east of Jensen Beach, a fragile strip of sand and concrete. Like many kids, our children have attached a sort of magic to their regular vacation spot. For them, Hutchinson Island is a carefree destination where they are doted upon by their parents and grandparents; a place to frolic on the beach, swim in the ocean, and occasionally venture out for a spring-training baseball game or a trip to the Everglades. At times, they have floated the notion of our family moving to Florida. And why wouldn't they? They likely imagine that such a life would mark an endless vacation free from school, chores, and snow boots. But what they have difficulty imagining is the prospect that someday, all that they love about Florida—those gentle coastlines, the fresh-squeezed orange juice, the largely languid seas—will be gone.

Back in Wisconsin, I listen to my children ramble on about how much they love Hutchinson Island, and what our life there could look like, before inevitably squashing their dreams.

"We'll never live in Florida," I say, pushing a wheelbarrow of firewood along our driveway.

"Why not?" they ask.

"Because it won't exist in fifty years."

This line of questioning isn't novel. It is something of a refrain, one that usually comes up just before or after vacation. But even though they fantasize about a Florida lifestyle, a part of them also seems to know, intuitively, why it might not be possible. From a young age, our children seemed to already know about concepts like *global warming*, *sea-level rise*, and *climate change*. Armed with facts they've likely learned from school and PBS cartoons, they are bright enough to hear my reasoning and connect the dots.

"Global warming?" our son will ask earnestly.

"Yes."

"So, Hutchinson Island will be gone?" Our daughter.

"Gone," I repeat. "Probably Jensen Beach, too."

I generally try to impart to my children a sense of wonder regarding the natural world, rather than the constant sense of impending danger and destruction. And yet, to be both realistic and scientific is to engage the horrifying facts. The natural world is under siege, and to ignore that reality is to do myself and my children a disservice. So, when they ask me about the plight of orangutans or polar bears or their beloved beaches, I don't lie to them. I can't count the number of years in which I have watched, with binoculars, ships suction sand a half mile off the Florida coast, only to redeposit it back onto hurricane-eroded beaches. Walking those shores, my wife and I pass huge pipelines that cough the sand up for its later redistribution by bulldozers. These are images I'd rather hide from my children. But they deserve to know.

"Miami?"

"Gone. Already going under."

"New York City?"

No doubt they are imagining a map of their known world, vectoring in tightly on blue coastlines and romantic cities they might have seen on family trips, others that they know from movies or TV or stories told by relatives. These are places I have explored in my lifetime but that may be gone in theirs. My world was bigger than the world they will inherit, a sobering thought for a parent focused on trying to ensure a better future for his children. The world I knew and explored was something that, of course, I want to share with my children—not just tell them stories of places that have disappeared.

I think of standing in the corner office in my publisher's headquarters, on the sixteenth floor of the Flatiron Building, and looking down upon Madison Square Park. Or the time, as a teenager, when my best friend and I drank beers in the Village Vanguard, or when we watched a live taping of the *Late Show with David Letterman* in the Ed Sullivan Theater. Drinking a cold glass of champagne with an old friend at the Met Roof Garden Bar. Watching a Yankees game in the Bronx back in 1988; my childhood hero, Don Mattingly, going 2-for-4 in old Yankee Stadium. Though I prefer living in the country, far from big cities, the notion of losing New York, a place that is so meaningful to me as an artist and writer, is breathtakingly tragic. And because of the city's image of invincible grit, I think we take for granted its fragility. We've seen the horror that two airplanes could inflict, or Hurricane Sandy, or the novel coronavirus pandemic—and now, on top of that, we must also imagine the incremental hunger of the sea?

"Unless they start moving fast," I say grimly, "they'll be underwater, too."

When I say *they*, I realize I mean millions of citizens, politicians, corporations, engineers, environmentalists, liberals and conservatives—everyone living in vulnerable areas across the United States. *They* includes hundreds of people I know well, some that I dearly love, many of whom I depend on. What does it mean that, in such moments, I think of the city as *they* and not *we*? And yet, I've also heard people talk about Wisconsin as if it were another country—How could *they* vote for him? Why do *they* need guns? Why do *they* always vote against their own interests? One reason we may never be able to solve this crisis in time is because we are so separated by tribe. Because we so willingly separate ourselves from one another.

I think of my father, who, all through my childhood, used to challenge me by saying, "What would you do if I wasn't here? How would you take care of yourself? Your mother? Your younger brother?" I never realized he was trying to prepare me for an unkind world—I thought he was just trying to discipline me—but I think I understand this parenting tactic now. When I talk to my children about climate change, I want to impart to them this sense of being aware, of being flexible. The world is constantly changing, and I want them to know how to adapt. The key, I think, to my own brand of parenting is to instill both a love of the natural world and the sober steadiness of science. In the coming decades, we'll need problem solvers—this is what I think my father was trying to teach me, and this in turn is what I'm trying to teach my children.

"Boston?"

Fenway Park, my best friend and me squeezed between two of my uncles after beers at the Green Dragon Tavern and a tour of the North Bennet Street School in Little Italy.

"Boston is built on dredged soil. It should already be underwater."

My children have never once asked me if we can stop the warming. They must intuit the answer from my tone, from the already dire state of affairs they've started to learn about in school and from our conversations.

Their very world is shrinking, and fast, as if the blue ink of every map were in fact still wet, still alive on the page, slowly spilling out to reclaim each green shoreline. Any road atlas from 1960, or political globe from 1900, will disprove the idea that a map is a static thing. The world is always changing. But now, instead of political boundaries being reshaped by invasion or social upheaval, the Earth is violently revolting against all of us, with huge implications for the physical borders of our world.

But it isn't just that the world is shrinking. The opportunities for memory-making are dwindling, too. The late afternoon strolls I have taken with my wife in Venice, walking hand in hand beside water so hungry for that fragile city. The memory of my friend and Italian editor, Patrizia Chendi, meeting us outside our canal-side hotel in a pair of knee-high Wellingtons as seawater lapped up over the stone walkways. In twenty years, what will be left of places like Venice for my children to explore? Or, much closer to home, how many farms will be left in the Wisconsin countryside when record-setting winters push farmers over the edge, both psychologically and financially? What will the identity of their home state look like when most of the farmers are gone? Even more troubling—where will their food come from?

During my childhood in the 1980s and '90s, I don't believe that the term *prepper* was in the popular lexicon—I'd certainly never heard it. But it would have been an appropriate thing to call my dad. My father collected guns—dozens of them. He hoarded ammunition, kept a supply of MREs, and took solace in heating our home with wood. And he loved the *Mad Max* movies.

The post-apocalyptic world of *Mad Max* is all sand and wasteland, no water—a vision of the future that may end up being pretty accurate. While much of the world will have too much water, not enough of it will be potable. Meanwhile, places like Australia, the setting of the films, will only get hotter and drier.

There is a character in *Mad Max 2: The Road Warrior* called the Feral Kid (played by a then eight-year-old Emil Minty). Against all odds, the Feral Kid survives in the Wasteland because of his ingenuity in the wilderness and his prowess with a deadly sharp boomerang.

My dad used to tell me, "You want to survive in this world, you need to be like that kid. There won't be any room for *feelings*, for being *sensitive*."

But I didn't want to be like the Feral Kid. I wanted to, and still want to, believe that the world is a good place, full of good-hearted people. I also knew, deep down, that I could never be the Feral Kid. I was a husky child with glasses who liked listening to Tchaikovsky on his Walkman, who liked reading books and organizing baseball cards.

I've never screened the *Mad Max* movies for my children. I don't care to show them these dark, fictionalized versions of what their world might look like—I prefer to answer their questions as honestly as I can. But I often think of what my dad said about what it took to survive in this world. Even now, watching my children happily play video games, or explore the borders of our land, I wonder if I am doing enough to prepare them for their future. Maybe I ought to introduce them to those apocalyptic visions of the future, as my dad did for me. Maybe I'm under-preparing my kids. Worse, maybe I'm in denial. But the reality is that there isn't a playbook for parenting in the apocalypse.

After the 2004 Indian Ocean tsunami that devastated Thailand, among other places, I had a difficult time feeling relaxed in

Florida. For me, the ocean had ceased to be a languid, photogenic plane of water. Instead, I could only imagine what I had seen on so many YouTube videos: that curious calm when the tide suddenly receded far from shore, leaving boats stranded on sand, where they had floated beside a pier only a moment earlier. The nonchalant narration of the amateur videographers. And then the horror in their voices when they realized that the crashing surf they saw *miles out* was actually a tsunami rushing back toward land.

On Hutchinson Island, I would watch my children frolic in the surf with a menacing anxiety, sure that this ocean, the Atlantic, would swallow my family at any moment. I imagined their colorful plastic buckets and shovels being sucked out in a chaos of flotsam. I devised plans I could enact for our quick escape. In truth, I understood that our only hope would be to climb to the seventh floor of our hotel and ensconce ourselves in a bathroom, away from any windows or glass doors. I visualized stuffing towels in the crack below the door to staunch any seawater. I worried that the 1970s-era cement building would not be unable to withstand such an assault.

It was either Jim Harrison or Ted Kooser who wrote: *On every topographic map / the fingerprints of God.*

After we moved to these sixteen acres, I bought a topographic map from the USGS that depicts the land upon which we live. The map hangs, framed, in a hallway outside the bedrooms where my children sleep. Looking at the southern portion, one sees a nearly nonexistent village called Cleghorn; little more than a bar, a milk-hauling business, some snowplows, and a defunct taxidermy shop bunched around a drowsy crossroads. But the map is dense with lines that show a steep and complex topography of ridges, hills, and draws. Run your finger northeast from Cleghorn about a mile, and you will find our property near a number: *1100*. Our elevation above sea level.

"You see that number?" I used to ask our kids.

They'd nod.

"The ocean will never touch us. Not here. You're safe here."

Even though my children know about the dire straits of our planet, I don't know that they spend much time worrying about it independently—a testament to the fact that I try to keep our household a very happy, bright, artistic place. But that doesn't mean that I don't worry.

My dad came of age during the Vietnam War. Though he was never deployed, he served in the Reserves, as a desk clerk, processing the wills and paperwork of soldiers. I understand now that he must have been riddled with survivor's guilt, and that he wanted to be ready, should anything like that ever happen again. My father was heavily involved with Scouting when I was a kid and was instrumental in ensuring that I received my Eagle Scout Award. Despite the fact that I showed no interest in being the Feral Kid, I think he saw Scouting as a way to prep me. The Boy Scout motto, after all, is *Be Prepared*.

And I can remember him scrutinizing every landscape as if searching for a natural foxhole or shooting lane. He saw the Earth as a fortress, something that could be used in his own defense and the defense of his family. As I grow older, and the planet grows warmer, I understand his anxiety.

•

I began writing this essay just after dawn, after our children had left for school. That was when the snow began falling. Four hours later, it is falling more heavily still. The visibility extends only a hundred yards or so.

As I watch the flakes continue to fall, the only solace I take is in the notion that, for now at least, the rotation of seasons is still predictable, and spring will surely come. This snow will all melt away, running down this ridge to join an ephemeral stream that connects to Lowes Creek, before feeding into the Chippewa River and thence into the Mississippi. All downhill from here.

Until this snow reaches the ocean.

SEASON OF SICKNESS

POROCHISTA KHAKPOUR

There was a time before all this, that much I know.

In December 2017, I went back home to Los Angeles to spend the holidays with my family. At that point I was the healthiest I'd been in a while; I ran a lot and did yoga daily. Prior to that, though, I had spent several years dealing with intense chronic illnesses—all stemming from my untreated late-stage Lyme disease. But I was sure, after years of intense conventional and experimental therapies, that I was past it. So I flew from New York to Los Angeles, as I often liked to do in the winters, to enjoy a few sunny weeks with my family.

Within the first week in L.A., I got sick. It took me a second to realize there was something off with the sky, a heavy haze, a bad smell. It was past October's peak wildfire season, so I didn't think of fires until I started seeing footage of what was happening just a few miles away from me. Passengers were driving on the 405 amidst what looked like a David Lynch montage: walls of ash and blaze all around the highway, pouring in from the Getty area. Little did I know that 2017 was going to be the worst wildfire season in California history up to that point. With a total of 9,133 fires burning and 1,381,405 acres of land affected, this year would become historic.[1]

As a child, I used to fear the Santa Ana winds. I remembered that old fear during that holiday season, when they were especially intense. Santa Anas have always been wildfire fuel, but this particular year, other weather conditions intensified their impact. According to the National Oceanic and Atmospheric Administration, 2017 was the third-warmest year on record for the United States, and it was the second-hottest in California.[2] Climate change was on everyone's lips, as it did not take a rocket scientist to understand how it contributed to the fires: the hotter the temperatures, the drier the vegetation, and the drier the vegetation, the more likely it is to burn. Add some Santa Anas to the mix and you had the proverbial fan for the flames.

Los Angeles was burning, Ventura was burning, and for some weeks it truly felt like we lived in hell. I would walk outside just to return back inside—the air was too hard to breathe. Soon, I started to have trouble breathing indoors. At first I was slow to understand that this could be a Lyme relapse, but it finally dawned on me: Lyme needed a trigger—in this case, poor air quality—and it thrived in poorly oxygenated bodies. Since book tour season would be upon me again in spring, I took action: my Christmas gift to myself was an Inogen G3 portable oxygen concentrator. It cost more than any used car I had ever bought, but it seemed important. At that point in my life, I had been in many ER rooms with Lyme relapses and knew oxygen was one of the only things that helped me. On and off for that entire season, I was connected by canula to my calmly purring device.

I bought a backpack for the device and my parents eyed it warily. Neither of them had ever heard of Lyme before I got diagnosed—my family were Iranian immigrants and Iran was not a Lyme hotspot. Neither, at that point, was California. My father thought my Lyme doctors were quacks, and my mother worried it was depression and anxiety that were really plaguing me, like they had plagued her at my age. After years of seeing me walk with a cane, the portable oxygen seemed too much for them.

"I will be fine," I kept saying. "I think the bad air just set a few things off. I am sure I am getting better!"

But I was not. Not yet. I did not get better until 2020.

On its website, the World Health Organization offers this warning: "Humans have known that climatic conditions affect epidemic diseases from long before the role of infectious agents was discovered, late in the nineteenth century. Roman aristocrats retreated to hill resorts each summer to avoid malaria. South Asians learnt early that, in high summer, strongly curried foods were less likely to cause diarrhea."

The WHO also warns of other climate-related effects. For example, global temperature increases of two to three degrees Celsius would increase the number of people at risk of malaria. The WHO also tells us that reforestation is linked to an increase in cases of Lyme, which has much in common with malaria. Wherever there are increases in tick populations, Lyme flourishes. But, of course, malaria itself was not gone even with so many medications and vaccines. An ABC News article from September 2019 claims that "malaria, a mosquito-borne illness, killed 438,000 in

2015 alone, according to the World Health Organization, and the insects carry Zika and West Nile viruses among others."

An article in the *New Scientist* also draws links between climate change and illness by showing how planetary warming is "making new areas into suitable homes for disease-carrying species" like in the example of Ebola. "For example, if the trees that fruit bats—believed to be a reservoir for the Ebola virus—rely on can grow in a new area, the bats can follow." Given that research on Lyme is still far behind where it should be, scientists and doctors have often followed models posed by these other infectious diseases that come from an overpopulation of hosts. After all, the ticks with Lyme cannot fly or jump—without hosts they cannot travel. Lyme-disease-carrying ticks were always "deer ticks" (also known as "blacklegged ticks") but we now know deer are not the only carriers: mice, squirrels, birds, and many other animals carry these Lyme-diseased ticks.[3] Just as climate change has affected vector-borne diseases like yellow fever, Dengue fever, malaria, and more, climate change has also dramatically altered the distribution of Lyme-disease vectors.[4]

It feels futile, almost embarrassing, to argue against the reality of climate change and its consequences. I grew up in Los Angeles, so I know smog, earthquakes, and wildfires well. But the nearly unprecedented wildfires in that late fall and early winter of 2017 were something I had never seen before. Everyone I knew said the same. We knew it was coming—we'd been warned for ages about global warming—but I guess we just did not know how soon.

And even as I watched a machine count my breaths as I

inhaled through a tube, I realized just how little I had understood how global warming could affect me personally.

In December 2017, when I left my Harlem apartment for that extended holiday in California, I had no idea it would be two and a half years until I found a home again. Within days of returning to New York, I felt toxic. I started having trouble breathing, then trouble swallowing. I felt horribly dehydrated no matter how much water I consumed. I had body aches and headaches, and I experienced excruciating neuropathy. Construction was being done on the apartment above mine, something that had not happened before in my time there. The building's management had been trying to push old Black families out, even though they were protected with rent control. The management succeeded with the family above me. And since this was the first time that unit had ever been touched, the construction workers had no idea what they were doing.

For weeks parts of my ceiling crumbled, sending clouds of toxic particulates into my apartment. I knew they contained tiny flecks of lead paint, and likely asbestos, too, but something I hadn't noticed yet was the presence of black mold—an environmental hazard that had likely been there for some time and was exacerbated by the construction. Eventually spots began to gather in my bathroom. By the middle of spring, I could barely breathe without cycles of panic attacks in my own apartment; I was on oxygen the whole time. I kept asking my landlord and super for help, but they kept denying I needed it.

My medical bills were in the high thousands, and exposure to the lead, asbestos, and mold was the proven factor behind my health collapse. My emails went from polite to desperate, with the possibility of death in between every line. *We are very lucky i did not die in there, which could have easily happened*, I wrote one day; *I continue to fight for my life*, I wrote on another; one email draft just had the subject header, *please help me*.

And so I began sleeping at the homes of various people in the city, from friends to absolute strangers I reached out to online. Not many people had a spare bed in New York City. I knew I could not stay there, so I eventually went back to California. But even though it was a comfort to see my family, the air out west was still making me sick, and by then I had become intensely allergic to mold. The unprecedented rains in California that spring—climate change leaving its mark again—had created mold problems in buildings that just weren't designed for moisture.

I decided to do what I had done years ago and return to Santa Fe, where I had a doctor I trusted from my previous time there in 2012, and a world of Lyme treatment. There were nutrient IVs to build my immune system, conventional pharmaceuticals like antifungals to address mycotoxin illness, plus a whole world of alternative resources, from my beloved apitherapy to ozone therapy. They were all extremely experimental and left of the mainstream in America, though my doctor reminded me that in Europe these were considered far more conventional among the general public. In Santa Fe, I had previously experienced such a miraculous comeback after several years of mysterious maladies that I believed recovery awaited me there once again.

But, of course, bad luck struck again: the city had also experienced a shocking break in its drought and was plagued with mold as a result. The adobe soaked up the moisture and mold was everywhere in the high desert. I moved from home to home, inn to inn, stranger to stranger. It was beyond draining, because I would hit dead ends everywhere—an Airbnb I first stayed at proved to have extra-strong VOC paints, which were impossible for me as I had become chemically sensitive; an inn just outside town was freezing cold and unbearably austere to the point where I could not fall asleep there; a friend's house would give me migraines so bad that they finally had it inspected and discovered mold; another friend's place made me feel very sick for reasons we could never figure out. Occasionally, deeply disturbed eccentrics would discover me from a Facebook post and take me in and within days I'd find out that they were dangerous racists. Nothing was right. Through this process, my body failed more and more: irregular heartbeat, shocking weight loss, air hunger, tremors, dizziness, dangerously low blood pressure, allergies to everything. I felt like my health had relapsed back to where it was before I ever got treated for Lyme.

"You're an environmental refugee," my doctor said, not realizing, of course, that my early years were devoted to being an actual refugee from war and revolution in Iran. "But unfortunately this state is overwhelmed with so many of you these days."

"Where can I go?" I kept asking him.

And more than once I saw him look at me with tears in his eyes. "I don't know."

My doctor was a compassionate man—equal parts conventionally trained MD and alternatively minded naturopath. He knew his patients far more deeply than most doctors did. And of course, he had saved my life in 2012. But I had to face that he had nothing for me now. He said I had CIRS (chronic inflammatory response syndrome), that mold illness was my main problem, and that it had retriggered my Lyme disease. Unless I could find a mold-free space to live in, none of his remedies could help me. His office and IV room were filled with others with environmental illness who could not find a "safe space" even in Santa Fe. It was clear that this was our future, that more and more, as our climate evolved, our collective health would suffer.

I briefly stayed with my old sensei in Santa Fe, a bee venom therapist who administered the apitherapy through the Japanese Hoshin method, but even her space made me feel allergic. Hives, trouble breathing, unbearable itching, throat swelling, sinus infections—I was miserable every moment I was awake, and when it was time to sleep, I was too uncomfortable to check out of consciousness. But my sensei understood. Climate change had affected her as well. She was distraught by the state of her hives, the bees unable to handle the dramatic shifts in temperature. Bee venom had once been a huge help for me, but now we both eyed the honeybees with pity. They could barely help themselves.

"What have we done?" she used to say, staring at the heavens for answers that never came.

It took years for me to get better.

Eventually I decided to go back to New York City. I stayed in a friend of a friend's apartment and hired a mold coach—a job I'd just learned about—to scout out an apartment for me. My mold coach had been homeless too and was very careful around certain neighborhoods in New York, constantly smelling "mold plumes" and "toxins" with different breezes, she claimed. Eventually I found a place in a suburban part of Queens—an old 1960 skyscraper that was built ruggedly and very clean, with doormen and elevators, and I set up my new life.

It wasn't as hard as you'd imagine for me to start over: by this point, I had lost all my possessions. Nearly half a year after vacating Harlem, the roof had finally collapsed on my old apartment before a friend of mine could salvage my books and clothes, the whole of the apartment overtaken by brown and black murk. I had also thrown out the few possessions I collected in L.A. and Santa Fe, as even those had been exposed to mold. I slowly bought a few things here and there, washed them in borax, and even bought a cheap portable infrared sauna online, which doctors claimed helped with detox from mold illness. I put two air purifiers in place; one cost $1,400. I went to get weekly "immunity" and "detox" IVs at a Lyme clinic, plus phosphatidylcholine IVs as part of a mold detox regimen. I took strong antifungals and settled into a routine. Very slowly I got better.

And I had to: I had lost a hundred grand, after losing what seemed like a hundred years to Lyme. I had to get well if only to work again, because I could not afford this illness. This whole time, I had no real income other than a crowdfund friends had set up. Thankfully, it had raised $60,000—mostly from fans of my writing from around the world, and many anonymous donors who likely had experienced this sort of misfortune themselves, as their unsigned notes seemed to indicate.

I used to look outside my window and think, When will I be well enough to join the world again? I never considered I would one day be sitting inside, looking out and wondering when the world would be well enough to be rejoined.

By February 2020, all my chronic aches and pains and problems were gone. I could function at full capacity again. I went on a trip to Europe in early March and managed without a cane. I felt like my old self again. I was well.

But when I returned home, the world was not.

Upon landing at JFK, I was shepherded to a special line, and this was a first: I had my temperature taken. I was prepared with several face masks and hand sanitizer, but I did not realize how precious those items would be in the weeks to come, or how hard it would be to get basic dry goods and toilet paper. The world turned upside down in a matter of weeks as COVID-19 spread not unlike the wildfires of my old home in Los Angeles.

The heart of New York's pandemic—of America's pandemic—was just a few blocks from my apartment in Queens. We had the highest number of cases in the country. News reports showed footage of a hospital in nearby Elmhurst, where bodies were being piled up in refrigerated trucks parked just outside the hospital's doors.

The panic people felt was new, but it also seemed familiar. Horror movies and dystopian novels have long painted a world ravaged by climate change disasters that looked a lot like this. And the connections between viral outbreaks and climate change were real. *The Washington Post* warned that climate change creates an opportune environment for the 320,000 viruses that infect mammals to thrive.[5] In *New York* magazine, David Wallace-Wells warned, "If the disease and our utter inability to respond to it terrifies you about our future staring down climate change, it should, not just as a 'fire drill' for climate change generally but as a test run for all the diseases that will be unleashed in the decades ahead by warming."[6] Meanwhile at *The Nation*, Ana María Archila, George Goehl, and Maurice Mitchell penned a piece arguing that our current crisis with COVID-19 could be a cautionary tale for our inevitable future with climate change: "Perhaps the most important lesson of the coronavirus is that if we don't prepare now, and start thinking about how to stop problems before it's too late, we're risking everything we care about: our homes, our jobs, and the health of our loved ones . . . Greater disease transmission, food shortages, energy blackouts, floods, homelessness, joblessness, species extinction—each will stagger us and then do so again."

Day after day I read articles like this, as New York made it through a winter with almost no snowfall, and into a spring that produced the worst allergies I can ever remember. The Kwanzan cherry trees in my neighborhood were almost steroidal in their unchallenged beauty this season. A frosty hailstorm hit just as weather forecasters were predicting a hot summer ahead, and then, on a gloomy day in April, a first for me: *a tornado warning in Manhattan.* My dog looked perturbed by the turbulent sky—he no longer wanted to run in the dog park.

•

It has been many weeks, and as I write this, we are still in COVID-19 season. As it turns out, you can get used to anything, even lockdown.

I still think about how grateful I am to have found a safe home at last, to have healed. But now I wonder what will get me in the end. *I was dying of America*, declared a scrawl I found in a journal from 2018 when I was at my absolute worst, suicidal in Los Angeles, years away from living even an hour without pain.

During the spring in which I write this, everyone I know has been taking photos of lush flowers and fertile trees, radiant blue skies. My parents text me frequently to report that the air in Los Angeles is more breathable than it has ever been—thanks to the COVID-19 outbreak, L.A. is seemingly overnight no longer one of the most polluted cities in the world. Even New York smells crisper, cleaner. I look out my window that faces Queens Boulevard and see just one or two cars speeding down eight lanes. The sirens of ambulances are constant—this sound, almost a song, I am entirely used to now. It mingles with the songs of birds. A couple of people in face masks walk their dogs. Signs have gone up to indicate buildings coming down, and more businesses have closed. So much is happening, so much is not happening. Sometimes I can't even remember what I miss anymore.

We know this pandemic will end and that it will take a long time, but probably the hardest thing we know is that this is just the first pandemic for us. There will be many to come and we will always be unprepared, or so we fear right now. *Nature hates us*, my friend jokes mock-bitterly, sending me a photo of her overgrown garden. But I think the truth is the opposite. All around us is potential and promise, from nature to science, waiting for us to take the cue to save ourselves. We wonder if our Earth's recovery will happen in our lifetime, but I think we know the answer.

Notes

1. Thomas W. Porter, Wade Crowfoot, and Gavin Newsom, *2017 Wildfire Activity Statistics* (Sacramento: California Department of Forestry and Fire Protection, April 2019), www.fire.ca.gov/media/10059/2017_redbook_final.pdf.
2. "2017 Was 3rd Warmest Year on Record for U.S.," National Oceanic and Atmospheric Administration, January 8, 2018, www.noaa.gov/news/2017-was-3rd-warmest-year-on-record-for-us.
3. "Birds Identified as Hosts of Lyme Disease," *Entomology Today*, March 10, 2015, entomologytoday.org/2015/03/10/birds-identified-as-hosts-of-lyme-disease.
4. "Climate Change and Vector-Borne Disease," UCAR Center for Science Education, accessed February 27, 2021, scied.ucar.edu/longcontent/climate-change-and-vector-borne-disease.
5. Sarah Kaplan, "Climate Change Affects Everything—Even the Coronavirus," *The Washington Post*, April 15, 2020, www.washingtonpost.com/climate-solutions/2020/04/15/climate-change-affects-everything-even-coronavirus.
6. David Wallace Wells, "The Coronavirus Is a Preview of Our Climate-Change Future," *Intelligencer*, April 8, 2020, nymag.com/intelligencer/2020/04/the-coronavirus-is-a-preview-of-our-climate-change-future.html.

THE DEVELOPMENT

ALEXANDRA KLEEMAN

I had lived in New York City for six years without a piece of nature that felt like my own, and then suddenly I had one: a long, narrow strip of land abutting the harbor, outside the Staten Island apartment building where I had recently moved from Brooklyn with my then boyfriend. The site was wrecked, abandoned, crowded with different kinds of brambles and leaves, mostly a dull and muddled greenish color. It was a patch of broken concrete shot through by stubborn weeds, some the size of small trees, and marked by eerie details: a fallen streetlamp, a collapsed pier. I walked alongside it most days on my way to the Staten Island commuter ferry, which I rode for twenty-five minutes to Manhattan, and then forty-five minutes more on the subway until I reached the school where I taught. Sometimes I paused on my walk to watch the tough brackish tufts sway in the wind, or I'd press my camera phone up against a gap in the chain-link fencing to take an unobstructed photo of the big cargo ships as they slid slowly out to sea.

As I passed by each day, I noticed new things: tall, stiff,

seedpod-topped grasses grew from the basin of a neglected landscaping planter, perhaps twenty feet wide—clearly, they had choked out the planter's previous occupant, some ornamental species less suited to wild winds and rains. Sometimes I saw fish fins dried out and stuck to the asphalt, or small lemon-yellow songbirds flitting from bush to bush. Buried in the thick brush were remnants of cultivated life: a rosebush nearly enveloped by overgrown juniper, its magenta blossoms mistakable for colorful trash; sprigs of irises growing tilted from a collapsing flower bed. Walking past here in both the daytime and at night, I saw with satisfaction that there was just as much happening in the darkness: unidentifiable scurrying, raccoons crawling out from the trash cans with affronted looks on their sharp faces. One night, under the too-bright cast of a streetlamp, I saw a scrap of red in the brush that made me gasp: as I came closer I saw it was an apple tree, bearing a full load of fruit in mid-November.

The overall effect was apocalyptic—but also beautiful. This strip of land, fifty feet wide at its broadest point and less than fifteen at its narrowest, contained more varied life than anyplace I had been in New York City—or maybe, over time, I had simply put in so much time watching that I was finally able to perceive it, like when you stare at a Magic Eye and suddenly the image appears, projecting from the background. I rarely witnessed anyone else stopping to peer into its disheveled green, to notice the ducks and geese paddling around the base of the blackened pilings. I never saw anybody photograph the night apples, or even recognize they were there. And though this made me feel more alone in the new place I had moved to, surrounded by so much to notice and no sense that we were all seeing the same things, it also made this odd, liminal place feel more specifically mine to love, mine for as long as it remained intact and above sea level. I didn't yet know that there was a plan to renovate the walkway, to replace what had been destroyed and make the area look safe and normal and whole once again. It was obvious that the sea would reclaim my little strip of nature eventually—but I hadn't thought to worry about developers claiming it first.

My partner and I had moved to Staten Island because we planned to move in together and thought we might not stay coupled in a Brooklyn railroad apartment, if we had to navigate together a space that was about as wide as he was tall. He had grown up on Staten Island in its suburban middle, and I came from a famously liberal college town in Colorado where you were never more than a few minutes' drive from a hiking trail. His descriptions of Staten Island read as exotic to me: Catholic school and Italian groceries, Sri Lankan brunch and disco roller-skating. On one of our earliest dates, a few months after Hurricane Sandy ravaged lower Brooklyn, much of coastal New Jersey, and the eastern shore of Staten Island, I asked him whether his parents had been on the island for the hurricane, whether they had been affected. He told me that they had spent Sandy sheltering calmly at home, but Hurricane Irene had hit them hard. He showed me a photo from years back of his mother in the local newspaper. She stood proud and smiling in their kitchen, posing beneath the trunk of a large tree that had crashed through their roof. The hurricane had caused major damage to the house, but nobody had been hurt. This, his mother had told him, was a sign from

God. Yes, the world was a dangerous place, but God would ensure that the faithful emerged unscathed.

If she saw the fallen tree as a sign of benevolent protection, I considered it an omen of catastrophic change to come. Hurricane after hurricane had hit Staten Island's shores over the years, and in the wake of Sandy's twelve-foot storm surges, a whole neighborhood, Oakwood Beach, was demolished, its residents offered a settlement in exchange for moving out of their homes.[1] Almost five hundred properties were returned to nature.[2] While I waited out the storm in a tiny room in Williamsburg, far enough inland that the power never stopped flowing, twenty-four lives were lost in Staten Island, most of them in the island's designated evacuation zones. I knew people in Manhattan or Brooklyn who had lost power and had been trapped downtown for days without cell service or access to public transportation. But here, the effects were much more catastrophic: more than half the deaths recorded in New York City as a result of the hurricane belonged to Staten Island. My partner's close friend from high school had evacuated with his wife and daughter and returned to find their home inundated. Two years later, we visited that same apartment, which they had renovated and waterproofed since the flood. The floor was covered in tile, the windows brand-new and designed to withstand heavy weather. Few things rested directly on the floor. The newness stood in for what had been destroyed, gesturing toward but not naming the disaster.

Knowing all this, it felt strange to sign the lease on an apartment that sat right on the crumbling waterfront—but the fairly low rent and fairly high ceilings were difficult to pass up. And on a sunny day, almost two years after Hurricane Sandy had hit the east coast, it was difficult to look at the sturdy, well-kept structure and imagine it buffeted by storms. The co-op building we moved into in St. George, a diverse and multicultural area on the North Shore, looked long-lasting. It inspired confidence because it had been around a long time: it was one of a set of converted warehouses built at the water's edge at the turn of the nineteenth century. (The building we moved into used to store coffee.) Large structural columns left over from when the space had been one continuous open area obtruded and interrupted the layouts of the subdivided units, and the windows looked out onto the sprawling parking lot and a small swatch of open water. Directly adjacent to our building was a company that ran tugboats for the harbor, cute dogged-looking crafts named after real people and bearing their full names painted on the side. In the other direction, a public square that looked populated only during the summer months, when fishermen with their poles came to dangle lines off the railing. After a year of sitting on the same furniture in the same arrangement, gathering my mail in the lobby, and staring out the window at the same quiet view, it became harder and harder to imagine this place as temporary or imperiled. Instead, it simply felt like home.

Time passed. We traded old jobs for new jobs, wrote books, got a dog, got married—and then, when an apartment on our floor went up for sale, we bought our first home together on the side of the building most often battered by the storms, the side with blinding morning sunlight and somber afternoon shadows. From our third-floor window, we looked down directly onto my beloved strip of greenery lining the walkway to the ferry terminal—and with a perfect axiomatic view, I could see into the deepest, densest parts, where feral cats slept beneath the rotting boards of wrecked boardwalk and one long pedestrian pier tilted precariously into the water, paused mid-fall. We built a long desk against one bank of windows and sometimes worked there side by side. I wrote parts of an apocalyptic novel while staring out at the planted pines and the resurgent shrubs whipping back and forth in the wind during heavy rains.

The lush, crumbling landscape was a useful prompt for thinking about deep time and not-so-distant futures: its slumped structures and eroding shapes made the clean-cut, fast-moving world of big-city fashion, culture, and finance feel a little bit like a lie. It opened up a space for imagining what might come next, what the world might look like as the water begins to rise up and claim its space, the world of the future that would replace our own.

Becoming a homeowner was like joining a secret club: our neighbors spoke to us more and told us unexpected things about where we lived. We learned that the homeowners' association had a policy of trapping and euthanizing the raccoons that lived under the abandoned dock. We learned who was a retired cop and who was a retired nurse, who was a Melville expert and who had voted for Trump. We learned that the entire first floor had flooded during Sandy, with water up to the thigh—and that the wrecked shoreline, hastily cordoned off by chain-link fencing, wasn't the work of Sandy, as I had assumed, but of Hurricane Irene a year earlier. This fact was alarming—at the same time, it seemed to alarm no one around me, at least not visibly. Maybe they had a head start on coming to terms with the precarity of our position, or maybe what seemed like apathy was actually acceptance, or resignation. It's difficult to advocate for a piece of land that nobody seems to care much about, and without the backing of developers drawn by the promise of rising property values and new, profitable projects, repair and reconstruction lack a sense of urgency.

Our building sits in stark contrast to the development in Brooklyn Bridge Park, which was hit by Sandy while still under construction. The park's design was shored up with features designed to resist similar weather events in the future: the grade of the land raised with geographical prosthetics, the edgelands planted with flood-resistant species. Soon after, the park was complete—a model of resilience, affluence, and forward-thinking care. In St. George, the waterfront has simply been fenced off for almost two decades, left to slump a few inches further into the harbor each year: the convergence of a complex engineering problem, low community engagement, and a piece of land whose value was difficult to define, difficult to leverage, and difficult to wring a profit from.

We also learned that the other members of our little complex of converted warehouses saw the damaged waterfront very differently from me—as absent of life, rather than serving as a refuge for it. Many of them were retired and had lived here since the 1980s or '90s: some remembered when it had been a well-maintained walkway, planned and landscaped with areas where you could sit and stare out at the water. Others remembered the recreation center that had stood there from 1934 to 2010, a 3.4-acre complex built out on a large wooden pier. Where damp-darkened pilings and open water now stood, people had played basketball and taken arts and crafts courses. The center had collapsed years ago, but a small security office still stood by its former entrance, slender young branches growing toward the sunlight from inside its weathered walls.

As a refugee from New York City neighborhoods that had gentrified in time-lapse around me, I found it refreshing to find a piece of land in the city that hadn't been capitalized on, hadn't been developed—a space whose exchange value and use value were compellingly obscure. Before moving to Staten Island, I had been pushed from one neighborhood to another by the economic churn of the city—small, homegrown eateries pushed out by snazzy, interior-designed restaurants, and long-occupied family buildings replaced by impossibly tall, impossibly narrow luxury condos. An empty lot allowed to remain empty felt, paradoxically, like a sign that I'd be able to stay. But to my neighbors, the damaged waterfront was a blight, suppressing property values—and now that I was a stakeholder, I was supposed to feel the same way, to root for the eradication of small mammals and look forward to the borough's plans, long-delayed, to rebuild the walkway and turn the crumbling concrete into crisp new pavement. It was true that the space was not pristine; you sometimes saw odd things in there that had been tossed over the fence—like a bowling ball or a brand-new, blindingly white pair of sneakers. But it was a special place, an unusual place, I had attached to it the way it was, and by the time the city announced that it was going to begin exploring how the area's renovation might actually be done, I no longer wanted it to change at all.

•

Slowly, gradually, I realized that the changes had already begun, were already in progress. Of course, I loved the abandoned land because it was honest about its own erosion, degradation, and transformation—about the fragility of seemingly permanent materials like concrete and stone. But at first it seemed like a transformation suspended in time, a permanent ruin, a handful of small seasonal shifts rotating around a steady axis. But after living here two years and then three, I saw exaggerated, ominous versions of the old things emerge, loudly announcing themselves as something new. During storms, the swell now came higher than ever, burying the tips of the pilings under gray water and sloshing over onto the walkway so that the water lapped at the asphalt the way the sea laps at sand. Now, after the storm, I detoured around large glassy stretches of water lingering on the hard surfaces—on the pathway and on the plaza and in the center of the communal lawn. I watched warily as new sinkholes formed in the parking lot, patched sloppily with loose tar and gravel or sometimes stuffed with a bright orange safety cone to warn us away. Last winter, a piece of my abandoned strip of land slid into the harbor, and now, at its narrowest point, the strip is only five feet wide.

Watching the accelerating disintegration of the waterfront made its quiet, melancholy beauty feel less meditative, less serene. When I looked out at the familiar scene, I was more likely to see something unfamiliar in it, an alarming new tilt to the landscape or a gaping hole. I found less solace, and more to worry over—but at the same time, how sad could I be about the loss of a place that I had always known was precarious? When I read news articles about hurricanes, flooding, sea level rise, I always tried to remind myself that these catastrophic changes were part of my home's destiny, that I would not be living in this apartment far into the indefinite future—or if I did end up doing so, it would be under radically different circumstances. But it was hard not to feel sentimental on a beautiful morning with the heat of the sun reflecting off the harbor's surface, hard not to feel attached when I thought about the holes we had patched by hand, the wall-paper we had worked all day to install. It was difficult to find the right amount of sentimentality to live with—too much and it felt like I lost touch with the shifting climate, the reality I knew was bearing down on us all, too little and it felt like I lost all sense of home. How can someone live in the present and the future at the exact same moment?

But the strangest thing about living near these changes is attempting to reconcile my view of them with those of my neighbors. On the one hand, nobody who owns property wants to imagine that their property is at risk: flooding is a disaster, but ultimately a random occurrence and one that can be repaired. But once repaired, it's a matter of chance whether you will be once again unluckily touched by disaster or luckily spared. To live within the basic precarity of life is one thing—to make your home on imperiled land is another. The first-floor residents who were flooded during Sandy sued the building, put their winnings toward new stormproof windows and kitchen floors, and stayed put. The co-op board voted to install new carpets and wallpaper and to try to host a movie night every four months. Walking to the ferry one rainy day, I pointed to the lapping water on the path and said to the woman walking next to me that by next year we might have to find another way to get there. Shaking her head, she replied, "They'll fix it, they'll fix it." A year later, the harbor still comes up to meet the asphalt, and even when the weather is dry the high tide swirls and retracts over the edge of the abandoned strip. When it rains, a pothole the length and width of a Jeep fills with water, and the next day a family of geese come to paddle gently around in it. I watch them from my third-floor window, and keep a count of the goslings.

A few months ago, the city came to test the structural integrity of the walkway, as it worked on putting together a plan for the waterfront's re-beautification. Week after week, men worked punching long, deep holes through the asphalt crust, and when they had done enough, they packed up and left. The holes are not so small: you could probably drop a baseball down a dark shaft, certainly a tennis ball. One day, while walking my dog along the waterfront, I stopped and crouched down, lowering my ear to the dark mouth of a hole, listening to the swish of unseen water moving down below. The plan for the revitalized waterfront involves building the walkway back out to its original, uneroded width, adding new ornamental plantings and stainless-steel benches. It's possible that in five or ten years, I could be leaning over the newly installed railing, peering out at the slow, sleepy procession of cargo ships, cruise liners, trash barges.

It's a possibility, but what seems more certain is that the sea will rise steadily, in accordance with the models of climate scientists, who warn of a 22 to 36 percent chance of a six-foot flood occurring by 2030. As you extend the model further into the future, that chance becomes 96 percent by the year 2100—essentially a certainty. A flood exceeding the nine-foot mark is equally likely within the same time span—and would be disastrous. A nine-foot flood would mean four or five feet of water in the first-floor lobby of my building, water submerging the wall of mailboxes and the glass case where movie nights are announced on home-printed fliers. Under nine feet of water, only the upper limbs of the scraggly apple tree would be visible—and most of the land-marks in my stretch of wrecked rewilding would be drowned.

I wonder what I would recognize of the area where I've lived for six years now—longer than I've lived anywhere else. Would the chain link topple or would I see the thin fringe of metal cutting through the water's surface? Would the cars parked in the lot out front go under, or would they float away? To imagine the destruction of a place I've spent so many beloved moments in is not only heartbreaking, it's physically difficult to do—even as I work to conjure the image of a wave sweeping in through my front door, the picture is eroded and replaced by mundane, typical scenes from my life here, the memory of normalcy etched so deep through repetition that it won't allow for deviation. The closest I can get is imagining a moment far after the disaster when the waters calm again, high and unreceded. From my third-floor window, you could watch the water lap against the flank of my building, licking at the storm-proofed windows on the ground floor. You could spot the gulls riding high on the gray-blue surface, the endless domain of water stretching all around and shifting in the light. All of these things could be seen from my window in a perfect axiomatic view—that is, if there was anyone still there to see them.

Notes

1. Sydney Kashiwagi, "State Purchased Hundreds of Sandy-Devastated Homes on Island, but One Homeowner Still in Limbo," *Staten Island Advance*, October 29, 2019, www.silive.com/news/2019/10/state-purchased-hundreds-of-sandy-devastated-homes-on-island-but-one-homeowner-still-in-limbo.html.
2. Ibid.

COUGAR

TERESE SVOBODA

It's four o'clock on a winter afternoon, nearly dusk. I'm halfway between Omaha and Lincoln en route to the airport. Here, for ten miles, taco joints and carpet stores give way to a stretch of farmland and woods. This all used to be farmland—no, this all used to be wilderness, the type of landscape you see now only in the small sections of cropland where the irrigation circles don't reach. But no snow anywhere. Unusual for this time of year. Except for the occasional terrible flood or tornado, the Midwest has few markers of climate change. Its farmers are always on the lookout for the slightest shifts—their livelihood depends on it—but their responses aren't always what you'd expect. For the last several years, my father, a Nebraska farmer for six decades, has shrugged off the lack of snow by telling me how higher carbon dioxide levels have increased his crop yield.

Recalling his enthusiasm, I'm roaring along in a rental car at around eighty-five miles per hour, ten over the speed limit, when something four-legged leaps across the two opposite lanes. I barely have time to slow down when I see it's a tan blip with an obelisk head and a tail nearly as long as its body. It tears across the median and past the front of my car, almost grazing my front bumper.

I park, shaken by the near collision. It's rare to see an animal at all near a highway these days, let alone a big one. Even roadkill and insects have become scarce. I peer past the windshield into the gray winter landscape, at the leaden clouds, the dead weeds, the leafless trees. I see nothing.

I pull out my phone to google what it might have been. A bobcat is too short to run parallel to the bumper, and they have "bobbed" tails, not the long apostrophe that streaked across my field of vision. I search for "cougar nearby." A month ago, a local kindergarten contacted the authorities to remove a female cougar that spent all morning sitting outside its front doors, seemingly waiting for children to appear. The local zoo was called in, and wildlife management loaded canisters of knock-out drugs into big bazooka-looking syringes, but the police were the ones who ultimately handled it. A photo of the kill shows two officers struggling to hoist the animal's 120 pounds.

Obviously, the cat should not have been hanging around the kindergarten, but it was probably just following its instincts—to be successful, a predator must be patient and ponder every move. My cougar didn't just jump into traffic. Unlike deer, which possess a startle reflex and will jump randomly to confuse predators, this cougar waited for a gap in traffic, took a chance, and barely cleared my car. I return my rental, thankfully intact, and fly home to New York City.

•

Big cats haunted me growing up in Nebraska, and they appear in more than one of my novels. When I was a child, local legend had it that they lurked in the limestone hills my siblings and I tobogganed down in the winter. Each time we turned our backs to mount our sleds, we expected to be pounced on—which made the kickoff even more exciting. Would a big cat leap on us on our way down the hill? We were always going too fast to check. We did see big animal prints while snowshoeing around the lakes in midwinter, though surely those were made by some plodding St. Bernard. Surely. As teenagers sitting around campfires redolent of sagebrush, we made up scary stories about big cats, the cinders glowing like the eyes of the cougars we imagined watching us from behind the spikes of yucca. Sometimes we would scream, not knowing that cougars scream too, rather than roar like lions. In the spring, after a local farmer spread tales about half-eaten calves, my father would laugh and say it was only a little bobcat. No need to worry.

But bobcat or cougar, if it had eaten one of his calves, my father would have killed it. He threatened to shoot whooping cranes if they so much as flew over his acreage, because, if their nests were found near Kingsley Dam, water authorities would halt the flow needed for local irrigation. He shot at prairie dogs because cattle broke their legs tripping on the burrowing rodents' holes, and he poisoned coyotes for bothering a colt. The pesticides he sprayed diminished huge flocks of birds, and most of the beloved pheasants that he liked to hunt in the fall. But he was not a raving homicidal farmer—all the other producers did the same. He was a soft-hearted bastard who brought home baby bunnies that he'd rescued from the combine; he sat up in bed and shouted in his sleep: "I'm going to kill you" over my cowering mother, then finished with "you broad-leafed plants." He must have inherited an unconscious drive from his pioneer forebears to conquer nature—or be conquered.

Due to unrestricted hunting by settlers (and consequent lack of prey), cougars had totally vanished from the Midwest by 1890. They were sighted again only a hundred years later. Once one of the most populous predators on Earth, with the greatest range of any terrestrial animal in the Western Hemisphere, the cougar spent the century holed up in a small area of southern Florida (as the panther) and in the western United States (as the "mountain lion" or "puma"). Between four and six thousand cougars still live in California, with one ensconced in the Hollywood Hills, having made its way from the Santa Monica Mountains through a corridor of tasty toy poodles and swimming pools and empty garages, and now six to fifteen breeding cougars have settled near my hometown in Nebraska. Another fifty-nine live three hours north. The cats are shy and elusive, but their return has not gone unnoticed.

"America's Cat on the Comeback" announced *American Scientist* in 2018.[1] Cougars are expanding their range, and even though humans tend not to like this, it's largely our fault it's happening. In the wetlands, the cats have been forced to move because of overdevelopment, and in the Midwest, they've lost their territory because of agriculture's tremendous growth. In 2018, when record-setting levels of ethanol production drove up the price of corn, farmers wanted to plant even the interstate's median strips to reap a greater profit.

To survive, cougars are seeking out green spaces wherever they can find them. In 2011 a male cougar was killed by a car in Connecticut. He had traveled all the way from the Black Hills, a trek of more than 1,800 miles. But at least the cougars aren't starving. Populations of prey species like deer and raccoons have reached a historic high. (Even in the middle of New York City, I have photographed raccoons on my back deck.)

Nebraska Game and Parks Commission, meanwhile, promotes the maintenance of "resilient, healthy, and socially acceptable populations" of cougar, with emphasis on "socially acceptable."[2] Few animals today are considered as such: cats, dogs, parakeets. The rest, if they're lucky, are not dead but miserable, behind bars. Cougars are the least aggressive of the world's large cats, but it's a tough sell to convince people that allowing cougars to run wild could be good for the environment. Truth is, they actually save lives. They hunt the deer that would otherwise step in front of your speeding car (the average mother cougar eats one deer every four days.) Cougars also diminish populations of prey animals, which in turn helps to rebalance ecosystems within their range. We've seen this happen before with other animal species. Wolves reintroduced to Yellowstone helped reduce the elk, which resulted in the return of foliage.

Yes, cougars may occasionally eat horses and cattle and toy poodles. And their size is terrifyingly formidable: they are bigger than wolves, with paws the size of a spread adult hand. But when it comes to humans, they keep their distance. It's statistically far more likely that you'd die from being attacked by a dog than a cougar.[3] They aren't entering our spaces because they want to threaten us—it's because we've left them no other choice.

•

The only cougar I encountered before my near miss on the interstate was Snagglepuss, the Hanna-Barbera Saturday morning cartoon cat from the sixties. An effeminate pink feline in an upturned collar who sported shirt cuffs and a string tie, he sounded a lot like Bert Lahr's Cowardly Lion: *I haven't any courage at all. I even scare myself,* although his most famous line was *Heavens to Murgatroyd!* Despite his lack of courage, the lion in Oz learned what it took to be brave. Snagglepuss had ambitions, but they always led him back to where he started and sometimes even worse off. His situation resembled that of the real-life wildcats: not quite protected, not yet extinct. In response to the cougar's comeback, thirteen states including Nebraska now advertise cougar hunts to get rid of "surplus" animals, without considering the competition of bears and coyotes, who often outhunt humans to kill cougar. In Saskatchewan, it's even legal to take cougars in bear traps, which causes the animals intense pain. If the traps aren't monitored on a regular basis, trapped animals die slowly; some even gnaw off their paws to escape.

Why do we need animals around anyway, aside from food or fur or cuddly comfort or anthropomorphized sex and death on TV nature series? After all, they spread viruses, they cause accidents, they bite. Even the activities we think they're vital for have workarounds: Man can always build robots to pollinate fruit and nut crops; Japan long ago produced a lovable mechanical dog to comfort its owners; and scientists have shown that the elderly appreciate the ever-attentive Alexa, whom I would classify as a pseudo-pet, given the affection she engenders. But what happened when China with its "Four Pests" campaign decided to eliminate rats, flies, mosquitoes, and sparrows? Severe famine. Animals, it turns out, are critically necessary to human life and civilization as we know it.

But even beyond the value of biodiversity, they warn us of environmental dangers. In 373 B.C.E., massive groups of rats, snakes, weasels, and other animals fled the Greek city of Helike days before an earthquake. In 2001, more than a dozen tagged blacktip sharks swam into deeper waters just before Tropical Storm Gabrielle decimated Florida's Terra Ceia Bay. In 2009, toads near L'Aquila, Italy, perhaps able to detect changes in the planet's atmospheric electrical fields, deserted their mating sites prior to an earthquake. The cougar is warning us. In the Midwest, its return announces that environmental pressures—resulting from climate change and habitat destruction—are so great that even large animals are on the move.

My two boys—men now—declare on Thanksgiving that we humans have at most only fifteen years left on this planet. I am so taken aback that I don't parry, and all of us go to bed in a panic. The sons are hardly doomsdayers: they have comfortable jobs in New York tech, and prepping connotes camping to them, which they abhor. But they've read the recent reports about the astonishing rate of global warming, the halving of the population of Earth's species, and this is their logical conclusion. The next morning, I sign everyone up for urban survival training, and we sit in frigid Central Park listening to the instructor review the contents of an emergency go bag, explain the importance of invisibility during escape, and teach us things like how to make fire and which edible plants grow in New York City. We have so many questions, the class goes on for five hours instead of the three we'd signed up for. I urge my sons to make friends with ten-year-olds, who will be stronger than they are in fifteen years. Afterward, we discuss where to flee—we're moving to Canada—and where we can plant trees next month. Where to put the bee house? Vote, I say, shower less. Learn how to make antibiotics so when the military rounds everyone up, you'll be more valuable. I tell them to do one thing about climate change every day, make a donation to an environmental group, water plants with cold shower water to reduce our resource use, install a solar panel on their tiny decks. How can I even consider having a family? asks the younger, whose metrosexual swagger barely conceals a sweetness no child can resist.

I have no answer for such despair.

My childhood Snagglepuss had a tendency to end every sentence with "even," as in "On account of I must be a little rusty. Stale, even," or "Somebody hurt! In dire pain, even!" a habit that intimates that he was expressing an extreme case of a more general proposition.

My sons, even.

They envision that we'll soon be like my highway cougar, running between corridors of disaster, searching for an environment that is not too cold or too hot or flooded or on fire or virus-prone. We'll be just like the animals, jiggling our feet over a rumbling disconsolate planet. What juggernauts of weather or earthquake will the luckiest of us just miss by a day or a mile? We pray for a freak rain to drench the Amazon, putting out a continent of fire, or a tsunami to judder to an end against an empty shoreline. While we still have a little time, perhaps like the cougar, we can leap clear of it. Avoid the utter devastation that climate change threatens.

Notes

1. Michelle LaRue, "America's Cat Is on the Comeback," *American Scientist* 106, no. 6 (November–December 2018): 352–59, www.americanscientist.org/article/americas-cat-is-on-the-comeback.
2. Sam Wilson, "Mountain Lions in Nebraska," Nebraska Game Parks, May 8, 2020, outdoornebraska.gov/mountainlions.
3. LaRue, "America's Cat."

SIGNS AND WONDERS

DELIA FALCONER

I'm walking to Mrs. Macquarie's Chair in Sydney's Domain at high tide, scanning the small bay in Woolloomooloo, as I always do, for fish or stingrays. There's nothing to see in the flat green water nudging the sandstone cliffs of the tiny beach, or below the seawall; I can't even spot the usual mullet nosing around the floating walkways at the marina. A few years ago, I might have assumed the variation in numbers was seasonal, hoping for better luck next time. But since 2016, when the figures started to come through that we have lost around 60 percent of the world's wildlife over the last half century—not only exotic animals but common creatures like giraffes, sparrows, and even insects—it's hard not to see today's emptiness as a sign of catastrophic absence.[1]

Things seem to carry a terrible freight these days. Swimming at Nielsen Park, in Sydney Harbor, an ancient river valley filled by melting Ice Age waters that stabilized six thousand years ago, I find myself wondering how high the water will rise again when the ice caps melt. "Every time I see a bird or bee these days," a friend says when we are talking on the phone, "I find myself wondering if it's the last."

The sense of loss is everywhere as each day brings news of unfolding disaster. Vanishing creatures are only part of a suite of ongoing catastrophes we are starting to recognize under the umbrella of the Anthropocene. Heating of the atmosphere and the rise of carbon dioxide, loss of forests, disruption of weather systems and sea currents, pollution from plastic and microplastics, and ocean acidification; together, these have accumulated the force of geological change, pushing us out of the stable patterns of the twelve-thousand-year-old Holocene and into a human-influenced new epoch.

And yet within the small span of one's own experience, it's hard to measure causes and effects, let alone how fast things are turning. As the world becomes more unstable in the grip of vast and all-pervasive change, it's difficult to discern exact chronologies, relationships, and meanings. In this unfolding context, small things take on terrifying and uncertain correlations. It's as if, I found myself thinking, as I scoured the water for fish, that in trying to see into the future we're returning to the dread speculation of the past. We've entered a new age of signs and wonders.

In ancient Rome, priests and officials called augurs would look for omens of the future in the weather, the movement of animals (especially animals encountered out of place), or the flights of birds. These days, we're scrutinizing the same things to tell the future, not as signs of the gods' will but our own actions.

Were the gale-force winds last November simply unseasonal,

I hear people asking in the playground, or evidence of global warming? Where are the summer cicadas, my mother asks—she can't remember hearing any on Sydney's north shore for years—and why have brush turkeys and rabbits started appearing in her garden?

Scientists are pursuing these questions with more rigor: modern augurs, staring at birds' intestines, are trying to read both the accumulated past and the future. Surely some of the most iconic images of the last decade have been Chris Jordan's photographs of dead albatross chicks on a beach on Midway Atoll National Wildlife Refuge on the Hawaiian Archipelago in the north Pacific Ocean.[2] In these pictures, which went viral, the birds' rib cages have collapsed to reveal, within the soft shape of decomposing feathers and bones, stomach cavities filled with dozens of bright pen lids, buttons, and bottle tops. The atoll is more than two thousand kilometers from the nearest continent. In a recent issue of *Griffith Review*, Cameron Muir describes watching scientists pump the stomachs of shearwater chicks emerging from their burrows for the first time on remote Lord Howe Island: the plastic objects they recovered—as many as 276 pieces per bird—confirmed the catastrophic spread of plastic pollution through the oceans.[3] Muir records that the belly of one young chick, which had to be euthanized, crunched in the scientists' hands. And yet, as seabirds decline faster than any other bird group, largely unseen, Muir writes, plastic production is likely to triple over the next thirty years.

Given the disconnect between what Muir calls the "shadow places," dead zones where we outsource pollution and disorder, and business as usual, it's hard to feel a smug sense of distance from the pre-scientific past, when people paid dread attention to omens. As we pay our mortgages, shop for clothes online, or plan our next holiday, signs of unfolding catastrophe are everywhere.

In the months I first began to think about this essay, there were out-of-control fires in tropical north Queensland and Tasmania's Ice Age World Heritage forests, and an "inland tsunami" when rains finally came to Queensland's drought-stricken Mount Isa. Unprecedented summer heat caused mass die-offs of flying fox colonies (up to a third of these native pollinators essentially "boiling," in a single record January week), and a million native fish in the lower reaches of Australia's largest river system, the Murray-Darling, an ecological disaster prepared for by the Murray-Darling Basin Authority's years of mismanagement and precipitated by extreme temperature fluctuations. (Since then, megafires in the summer of 2019–2020 have consumed most of the eastern seaboard, killing an estimated 3 billion animals.)[4] Meanwhile, in the freezer of my inner-city apartment, there is still a bucket of fist-sized, cauliflower-shaped hail, which I collected with my children as a "once-in-a-lifetime" hailstorm battered Sydney and its north coast just before Christmas.

These events have occurred, in turn, against almost daily announcements of global apocalypse (most recently, of a developing mass insect extinction or "insectageddon"; the accelerating melt rate of Antarctica's 34-million-year-old ice; the inevitable "doom" of a third of the Himalayan ice cap upon which almost 2 billion people depend; and that we are already in a period of contraction as extreme weather is shrinking the habitable spaces of Earth). At the same time there is a queasy uncertainty about cascades and tipping points.

All of these signs feel allegorical and at the same time they don't. They are so overwhelming, and so interlinked, that it is almost impossible to think outside them, to see beyond a pervasive dread.

Then there are the wonders, less obviously urgent, but pernicious in their atomized, quasi-magical effects. At the same time as global disarrangement is giving birth to signs of distress, it is throwing up phenomena of spectacular and haunting strangeness. Beautiful and uncanny phenomena are passing through our virtual atmosphere—through Facebook, Twitter, and Instagram—trailing the weird incandescence that must once have attached to Halley's comet or the northern lights. I became so fascinated by the eerie phenomena appearing almost daily in my feeds that I started to record them.

Most miraculous, surely, are the images of long-extinct animals emerging after tens of thousands of years from melting permafrost. In Canada's Yukon, gold miners recently unearthed a prehistoric wolf cub, its fur, skin, and muscles perfectly preserved.[5] Eight weeks old when it died fifty thousand years ago, when this forest was treeless tundra, the cub is the only one of its kind so far to appear; and yet, laid out on surgical gauze on a desktop in the official photograph, with its thick honey-colored fur, creased muzzle, and long, closed eyelids, it could be the Instagram photo of someone's sleeping puppy. That same year, from the Batagaika crater in Yakutia, in Siberia, a two-month-old Lenskaya, or Lena Horse, emerged from a sleep of thirty to forty thousand years, along with a month-old cave lion cub, found with its head resting on its paw, having died before its eyes had even opened.[6] These creatures have a corollary in the Russian deep-sea fisherman Roman Fedortsov's popular Twitter account featuring photographs of bizarre bycatch, including living "relics" such as the eel-like frilled shark, which have been trawled up from the mesopelagic or "twilight" zone of the Barents Sea. (It goes unremarked in the numerous aggregator sites and retweets of these pictures that the expanding bottom-trawl-fishing industry presents a major threat to global deep-sea biodiversity.)[7]

Throughout 2018, the world's fourth-hottest year on record at the time, global warming caused other ancient things to make themselves known in the northern hemisphere summer. Severe drought in Europe, in which rainfall in some places was 3 percent of the usual quota, saw a German river disgorge unexploded bombs from the Second World War. Archaeologists in Norway recovered hundreds of objects up to 1,500 years old from a lost Viking "highway" in a melting (and now entirely melted) mountain ice patch, while, in nearby Sweden, a girl pulled a pre-Viking-era sword from an "extremely low" drought-afflicted lake.[8]

Meanwhile, in the United Kingdom, the footprints of vanished Roman mansions, airfields, Victorian mansions, and prehistoric settlements were manifesting in grassed fields and parks. Straw yellow, or lush emerald green on lighter green, these eerie patterns were "parch marks": ghostly scars of human activity that reveal themselves as the land dries and grasses die off. In one haunting image, taken above farmland in Eynsham, Oxfordshire, a "harvest" of darkly outlined Neolithic barrow graves and paths and walls fills two vast fields with the inscrutable ceremonial structures of a lost society, dwarfing a modern farmhouse tucked into its tiny patch of garden. The dark green circles, lines, and smooth-edged squares made the yellowing fields look disconcertingly like the pages of Leonardo da Vinci's notebooks, as if a giant hand had made busy calculations across the earth itself.

There was actual writing, too, in the form of the dozen "hunger stones" that emerged from the drought-stricken Elbe River, near Děčín, in the Czech Republic, recording low water levels caused by "megadroughts" dating back as far as 1417. The inscriptions, in German, are also warnings to future generations. "Wenn du mich siehst, dann weine," one from 1616 read: If you see me, weep.

"If you see me, weep" seems like pertinent advice for living in the Anthropocene: this new geological era in which human activity has become the dominant influence on climate and environment and species loss is occurring, fifty years into the "Great Acceleration," at a thousand times the normal background rate. It is hard, in the middle of all this uncertainty, to shake the feeling, which the Eynsham parch marks produce so strongly, that these phenomena are also trying in some way to talk to us. In the Batagaika crater, a "megaslump" in the Siberian wilderness a kilometer long and almost eight hundred meters wide, the sound of running water and chunks of frozen ice thumping down the cliffs from the unstable rim announces the rapid melting of ground frozen for thousands of years. "As you stand inside the slump on soft piles of soil," one ecologist told *The Siberian Times*, "you hear it 'talking to you,' with the cracking sound of ice and a non-stop monotonous gurgling of little springs and rivers of water."[9]

The ecologist's description is reminiscent of the "general burst of terrific grandeur" described in Edgar Allan Poe's "A Descent into the Maelström," and it is almost as incredible.[10] The cheerful rescuers in Poe's story dismiss the mariner's tale of surviving the maelstrom as the fancy of an overactive mind. Not so long ago, it would have been possible to also dismiss the ecologist's observation as an instance of the "uncanny," defined by Sigmund Freud in 1919 as a dread and creeping horror that occurs when the hidden or secret seems to become visible and something once very familiar acquires an eerie sensation of animation.[11] Such instances, Freud wrote, "force upon us the idea of something fateful and inescapable where otherwise we should have spoken of 'chance' only."[12]

Freud considered such instances as throwbacks to an old, animistic conception of the universe—but the French philosopher Bruno Latour suggests the Earth really is speaking. Going back to Michel Serres's 1990 work, *The Natural Contract*, he repeats the assertion that the Earth is no longer the distant, objective foundation of our lives but now is so entangled with us that it is unstable and "trembling."[13] Once we accept that the fragile Earth is no longer a place of objective facts and, further, that our human activity is present everywhere, Latour argues, the world becomes "an active, local, limited, sensitive, fragile, quaking, and easily tickled envelope."[14]

Reading Latour, I find myself thinking of humoral theory, that ancient belief, which survived into the nineteenth century, that we are made up of four humors (blood, black bile, yellow bile, and phlegm). These corresponded, in turn, to the common elements (air, earth, fire, and water) and the movements of the planets: a theory that no longer seems quite so quaint or distant. And yet the expressive world that Latour envisages is also entirely different, because instead of encountering fixed laws, we are instead coming face to face, as our very existence hangs in the balance, with natural forces pushed past their tipping point by us to speak—like those oracles of old—with a terrible animation and agency.

I'm not sure that even Latour's writing quite captures the weirdness of our moment, in which time itself seems out of joint. When I think of the Batagaika crater giving birth to strangely uncorrupt young animals preserved for tens of thousands of years, I'm reminded of John Carpenter's horror film *The Thing*, set on a research center in the Antarctic, in which a parasitic entity that can perfectly imitate other creatures is inadvertently released by human scientists from a hundred-thousand-year-old alien shipwreck. Every part of the "thing" is an individual life-form with its own survival instinct, and until the scientists find a way to kill it, it threatens to assimilate all life on Earth. In the film's most memorable scene, as the creature is in its death throes, it morphs in a rapid, terrifying sequence into every animal and human whose life and form it has absorbed. We're experiencing as Gothic a haunting now, you could argue, by our fossil fuels. For years human life has run on extracting and burning hydrocarbons from the remains of decayed plants and animals—some more than 650 million years old—which are returning to ghastly life now in the form of global warming and its effects.

The implications of all these "wonders" are truly horrifying. A past that long precedes us—like Fedortsov's huge-eyed ghost sharks and rat fish—is appearing at its most vivid and strange as the conditions of its existence fall away. Yet these phenomena appear, in the media or in our feeds, as discrete objects of amusement and wonder. "Something about this reminds us of The Shape of Water," jokes an anonymous writer on a CBS photo gallery in response to a "crazy-looking fish" in Fedortsov's feed. "This looks like my sister," remarks an Instagram user in response to another. Yet we have only explored less than 0.05 percent of the "twilight" zone, whose creatures scientists believe may "pump" carbon from the surface to the seabed, even as fishing nations begin to exploit its immense masses of pelagic shrimp as feed for farm-raised fish. Or take a recent piece in *The New Yorker* on parch marks. After a paragraph paying lip service to the weird awfulness of the unusual hot weather, the writer quickly shifts his attention to the bonanza these marks represent for aerial archaeologists. "It's a bit like kids in a candy shop," one says, as the article goes on to speak to other excited beneficiaries of these "freak conditions."

Our huge appetite for such images, isolated in their own strangeness, makes me suspicious of the call from some environmental writers, like George Monbiot, for an increased sense of "wonder" as a way of saving the world by countering the sanitizing numbness of scientific language. It seems to me that the web is already a virtual cabinet of curiosities, inviting us to marvel—and yet our wonder rarely translates into action. Is wonder itself a kind of self-administered anesthetic, a means of telling ourselves that what we are witnessing is exceptional, rather than the rule? At the same time, we've become so quickly habituated to such sights, separated from their terrifying contexts, that it's hard to know if we're even in the territory of the "uncanny" anymore. As Zadie Smith has written in an "elegy" for England's seasons, the unfamiliar has become so pervasive that it is now "our new normal."[15]

These objects amuse and perhaps even mesmerize us. But where is the despair, the terror, the justified outrage?

The weirdest thing about our modern era, as Latour insists, is our adamant refusal to hear the Earth speaking as it takes on more capricious agency. And yet it is also hard to know how to feel in the face of these overwhelming signs and wonders. We are "not equipped," even Latour admits, "with the mental and emotional repertoire to deal with such a vast scale of events" or the new "emotions" that the Earth, disturbed by us, is expressing.

Even with so many signs within plain view, to even acknowledge the scale of this distress can feel unhinged, as if one is embracing the mutterings of a Nostradamus or the paranoia of a doomsday cult. "Have you gone through the terrible guilt yet about having children?" a colleague asks me in the corridor at work. Yes, I have, I answer. But she and I keep our voices low, beneath the fluorescent lights.

I can't help feeling that Freud, in spite of my great fondness for him, has a lot to answer for, in making us keep our voices hushed. It's hard to articulate our sense of the world's distress, when for so many of us our sense of modernity is founded on our distance from our animistic roots. Any person who had "fully banished animism from his soul," Freud claimed, would not be susceptible to the uncanny.

For Freud, the feeling of uncanniness—when "something we have hitherto regarded as imaginary appears before us in reality"—was a hangover from "our primitive forefathers," whose animism "civilized people" had surmounted. Beliefs in things like the return of the dead or the strange animation of the inanimate were also related, he claimed, to infantile impulses toward wish fulfilment that adults should normally overcome.

This means that while we may be prepared to address individual symptoms of global warming—to try to save a bird colony or old-growth forest—responding with urgency to the multiple signs of a world in distress can be easily dismissed as irrational or flaky. The 2018 report by the UN Intergovernmental Panel on Climate Change on the potential impacts of global warming of 1.5 degrees Celcius or more, and its prediction of a frighteningly small window of time (twelve years) left to limit global catastrophe, is one of the most galvanizing calls to action issued to humans so far.[16] Yet even it has been hampered, according to some critics, by its hesitation to name the "known unknowns" of climate change: tipping points or feedback mechanisms, which are the cascading and unpredictable outcomes of an already-dire set of threats.

I had forgotten, until I reread "The Uncanny," that Freud makes a distinction toward its end between experiences of the uncanny in real life, which in his own experience were rare, and their appearance in fiction, which he generally found more affecting. Yet these days, the ratio seems reversed. As the novelist Amitav Ghosh has argued in *The Great Derangement*, we are more likely to confront signs of damaged nature we have been trying to ignore forcing itself into our everyday reality—like those uneasy discussions I've been having on bus stops and in the playground—than in fiction, which eschews them as sensational or contrived.[17] Meanwhile, scientists such as Robert Larter, of the British Antarctic Survey, are using terms from fairy tales like "sleeping giants" to describe polar ice sheets, with their capacity for devastation.

Many novelists are nevertheless hard at work, also trying to bring our era's buried "structures of feeling" to light: writers of "weird" fiction like Jeff VanderMeer, and the long-view novel, like James Bradley, but also chroniclers of the present like the Scottish novelist Ali Smith. *Winter*—the second book in her "seasonal" quartet—opens with a retired businesswoman going about her chores in London as she is haunted by a child's floating head. "Bashful in its ceremoniousness," the head bobs and nods merrily in the air next to her "like a little green buoy in untroubled water."[18] For my money Smith comes closest, as a novelist, to invoking the stupefying quality of wonder in our present. The beatific head, which grows then begins to age, is in some ways less strange than the rituals and incantations of the neoliberal economy that oppress the heroine, Sophia (the two, she suggests slyly, are symbiotic). Through Sophia, she also channels the bursts of impotent anger the pervasive uncanny sometimes provokes, only for us to direct our feelings at nearby objects rather than root causes. But even Smith's quartet stops short of a galvanizing rage commensurate to a new reality in which "permanently wet" Gondwana forests on Australia's east coast (and the most ancient songbirds in the world) are being incinerated for the first time while gases escaping from melting permafrost are making the Arctic ground erupt "like a bottle of champagne."[19]

Yet I have to confess that seeing the world as possessing agency has never seemed that strange to me. That's partly my own nature, and partly because I came of age in Australia with some small sense of how Indigenous law, or lore, offers a powerful rebuff to Freud and our deafness to the Earth. Recently, Indigenous Australian scholars like the Dharawal elder Fran Bodkin and the Bunurong author Bruce Pascoe have written about the long histories of Indigenous meteorology and agriculture, which involve a contingent reading of natural cycles and behavior, always bound by a duty of care toward country in which culture and nature are linked.[20] Pascoe, the author of *Dark Emu*, has teamed up with Indigenous Australians along the New South Wales South Coast and in East Gippsland, in Victoria, to trial growing native plants such as kangaroo grass and murnong (yam daisy), which Indigenous people nurtured with complex ancient farming techniques and milled into bread.[21] In her last work on flying foxes living on the edge of extinction, published posthumously,[22] the late American-Australian environmental humanities scholar Deborah Bird Rose was working on building a bridge between extinction studies, with its emphasis on the complexity of relationships in landscape, and the Yolngu concept of *bir'yun*, or "shimmer," in which the world is composed of complex, multispecies "relations and pulses."[23] Compared to this way of understanding the world, which survived the last ice melt, even Serres's and Latour's analyses start to look shallow.

Meanwhile, as science advances, so much of what it once viewed as imaginary only continues to become real. At this moment, when they are most under assault, we are beginning to understand, in scientific terms, the extraordinary interconnections of nature's systems. Scientists have discovered, for example, that trees communicate and share food via mycorrhizal networks:

the "secret" life of trees has captured the public imagination in the form of numerous articles on the "wood wide web." Even our own bodies, it turns out, are only roughly half human, the other half made up of about 160 different bacterial genomes. Developmental biologists like Scott F. Gilbert are even suggesting that we consider ourselves as "holobionts": combinations of host and microbial community.[24] Next to this, economic models of "management" and sustainability look more and more like magic thinking. What use will our seed banks be without companion species of plants or pollinators, whose relationships have developed over almost unthinkable periods of time?

The challenge ahead is not only to accept that it is rational to understand that we are in the grips of a maelstrom of our own making—but also to let ourselves feel fear, awe, and rage equivalent to the "terrific grandeur" of nature out of whack. Replacing fossil fuels is the first urgent step. But allowing ourselves to be galvanized by genuine Edgar Allan Poe–level terror—those "unmodern" feelings we have shunned—at the signs of human activity reflected and distorted back at us, wherever we look, is what might save us all, human and non-human, in the long run.

Still, where do these thoughts leave me, on a bright blue day, as rainbow lorikeets and corellas call from the park below my study window, and I get ready to pick up my twins from school? How do I answer my seven-year-old daughter when she asks, unprompted, "Is it true that the world's going to end soon?"

I'd like to think that it's possible, in the staggeringly tiny window left to us, to act urgently: to see nature, as the anthropologist and writer Michael Taussig puts it, "not . . . as the dead, soulless object of European modernity but as something roused into life through the wounds and war conducted against it."[25] But at the moment, I just can't get past the grief—the sense that it's too late. I think of Poe again and the raven's call: *Nevermore.*

As I sit at my desk, an email comes through from my neighbor about the renovation works in the apartment block next to ours, which have involved digging down to its piping. Just by the way, she signs off, have I noticed the dearth of insects? At this time of year we would usually expect an invasion of spiders and moths through our unscreened windows. Is it because they've dug up the whole garden, or is this a sign of the insect apocalypse?

"I actually miss the huntsmen," she writes, "and even the thrilling and terrifying spider wasp. Our little biosphere, gone."

Notes

1. Monique Grooten and Rosamunde Almond, eds., *Living Planet Report—2018: Aiming Higher* (Gland, Switzerland: World Wildlife Federation, 2018), www.worldwildlife.org/pages/living-planet-report-2018. This 2018 report estimated that world wildlife populations had declined by 60 percent between 1970 and 2014. This updated the previous report, published in 2014, which estimated a loss of 50 percent. Rates of decline are worst in the tropics, with declines in Central and South America estimated at 83 percent.
2. Chris Jordan, "Midway: Message from the Gyre," *New York Review of Books*, November 11, 2009, www.nybooks.com/daily/2009/11/11/chris-jordan.

3. Cameron Muir, "Ghost Species and Shadow Places," *Griffith Review* 63: Writing the Country.
4. Graham Redfearn and Adam Morton, "Almost 3 Billion Animals Affected by Australian Bushfires, Report Shows," *The Guardian* July 28, 2020.
5. Meilan Solly, "Gold Miners Unearth 50,000-Year-Old Caribou Calf, Wolf Pup from Canadian Permafrost," *Smithsonian Magazine*, September 17, 2018, www.smithsonianmag.com/smart-news/canadian-permafrost-yields-intact-remains-50000-year-old-caribou-calf-wolf-pup-180970301.
6. James Gorman, "A Wolf Pup Mummy from the Ancient Arctic," *The New York Times*, December 21, 2020; "Perfectly-Preserved Ancient Foal Is Shown to the World for the First Time," *Siberian Times*, August 23, 2018, siberiantimes.com/other/others/news/perfectly-preserved-ancient-foal-is-shown-to-the-world-for-the-first-time.
7. Antonio Pusceddu et al., "Chronic and Intensive Bottom Trawling Impairs Deep-Sea Biodiversity and Ecosystem Functioning," *Proceedings of the National Academy of Sciences of the United States of America*, June 17, 2014;111(24):8861-6. doi: 10.1073/pnas.1405454111. Epub 2014 May 19.
8. Erin Blakemore, "Lost Viking 'Highway' Revealed by Melting Ice," *National Geographic* April 16, 2020, and (no author), "Girl, 8, Pulls a 1500-Year-Old Sword from a Lake in Sweden," BBC News website, October 4, 2018, www.bbc.com/news/world-europe-45753455.
9. "Disturbing Melody of Melting Permafrost in 'Crater' Called 'Gateway to Hell,'" *Siberian Times*, March 10, 2017, siberiantimes.com/other/others/news/n0895-disturbing-melody-of-melting-permafrost-in-crater-called-gateway-to-hell.

10. Edgar Allan Poe, "A Descent into the Maelström," in *Complete Stories and Poems of Edgar Allan Poe* (New York: Doubleday, 1966).
11. Sigmund Freud, "'The Uncanny' [1919]," in *Standard Edition of the Complete Psychological Works of Sigmund Freud*, ed. James Strachey, vol. XVII (1917–1919); *An Infantile Neurosis and Other Works* (London: Hogarth Press, 1955), 217–56.
12. Freud, "The Uncanny."
13. Michel Serres, *The Natural Contract*, trans. Elizabeth MacArthur and William Paulson (Ann Arbor: University of Michigan Press, 1995).
14. Bruno Latour, "Agency at the Time of the Anthropocene," *New Literary History* 45, no. 1 (2014): 1–18.
15. Zadie Smith, "Elegy for a Country's Seasons," *New York Review of Books*, April 3, 2014, www.nybooks.com/articles/2014/04/03/elegy-countrys-seasons.
16. Intergovernmental Panel on Climate Change, Global Warming of 1.5°, October 6, 2018, www.ipcc.ch/sr15/. In 2019, there was a growing consensus that this window was probably only eighteen months. See Matt McGrath, "Climate Change: 12 Years to Save the Planet? Make That 18 Months," BBC News website, July 24, 2019, www.bbc.com/news/science-environment-48964736.
17. Amitav Ghosh, *The Great Derangement: Climate Change and the Unthinkable* (Chicago: University of Chicago Press, 2016).
18. Ali Smith, *Winter* (London: Hamish Hamilton, 2017).
19. Ann Arnold, "Bushfires Devastate Rare and Enchanting Wildlife as 'Permanently Wet' Forests Burn for the First Time," ABC News website, November 27, 2019, www.abc.net.au/news/2019-11-27/bushfires-devastate-ancient-forests-and-rare-wildlife/11733956, and Andrew E. Kramer, "Land in Russia's Arctic Blows 'Like a Bottle of Champagne,'" *New York Times*, September 5, 2020.

20. Frances Bodkin, *D'harawal: Seasons and Climatic Cycles* (Sydney: F. Bodkin and L. Robertson, 2008).
21. Bruce Pascoe, *Dark Emu* (Broome, Western Australia: Magabala Books, 2014).
22. Deborah Bird Rose, *Shimmer: Flying Fox Exuberance in Worlds of Peril* (Edinburgh: Edinburgh University Press, 2021).
23. Deborah Bird Rose, "Flying Foxes on My Mind," Deborah Bird Rose: Love at the Edge of Extinction, November 23, 2018, archived June 20, 2019, web.archive.org/web/20190620222736/http://deborahbirdrose.com/tag/flying-foxes-2.
24. Scott F. Gilbert, "Holobiont by Birth: Multilineage Individuals as the Concretion of Cooperative Processes," in *Arts of Living on a Damaged Planet: Ghosts and Monsters of the Anthropocene*, ed. Anna Lowenhaupt Tsing et al. (Minneapolis: University of Minnesota Press, 2017), M73–M90.
25. Michael Taussig, *Palma Africana* (Chicago: University of Chicago Press, 2018), 104.

ACKNOWLEDGMENTS

We are grateful to our contributors for their beautiful essays. Much of their writing and revision took place during the pandemic, and we appreciate them persevering alongside us through unimaginable circumstances. This book would not be possible without them.

To our agent, Rayhané Sanders, who was excited about this project from the beginning: thank you for providing invaluable support at every stage, including helping us assemble our lineup of contributors.

Our deepest thanks to the entire Catapult team, including Nicole Caputo, our cover designer; Colin Legerton, our eagle-eyed copy editor; Selihah White, our publicist; and especially our editor, Leigh Newman, for being this project's champion and sounding board. We're so happy to call Catapult home.

Finally, to our extraordinary friends, families, and partners, Alan Scherstuhl and Philip Sayers, whose love and encouragement carried us through the hardest of days: thank you.

TEXT PERMISSIONS

"Starshift" by Gabrielle Bellot
A version of this essay appeared in *Guernica* (2019)

"What We Don't Talk About When We Talk About Antarctica" by Elizabeth Rush
A version of this essay appeared in *Literary Hub* (2019)

"Faster Than We Thought" by Omar El Akkad
A version of this essay appeared in *Literary Hub* (2019)

"After the Storm" by Mary Annaïse Heglar
A version of this essay appeared in *Guernica* (2019)

"Moments of Being" by Kim Stanley Robinson
From *The High Sierra* by Kim Stanley Robinson, copyright © 2022. Reprinted by permission of Little, Brown, an imprint of Hachette Book Group, Inc.

"Signs and Wonders" by Delia Falconer
A version of this essay appeared in *Sydney Review of Books* (2019)

ABOUT THE CONTRIBUTORS

GABRIELLE BELLOT is a staff writer for *Literary Hub*. She grew up in the Commonwealth of Dominica. Her work has appeared in *The New Yorker*, *The New York Times*, *The Atlantic*, *The Guardian*, *Shondaland*, *Guernica*, *Slate*, *Tin House*, *The Paris Review Daily*, *Los Angeles Review of Books*, *The Cut*, *Vice*, *Electric Literature*, *The Normal School*, *The Toast*, *TOR.com*, *Caribbean Review of Books*, *Small Axe*, *Autostraddle*, and many other places. She is the recipient of the 2016 Poynter Fellowship from Yale and also holds a Legacy Fellowship from Florida State University. Bellot holds both an MFA (2012) and a PhD (2017) in Fiction from Florida State University, and currently teaches classes at Catapult and Gotham Writers Workshop.

NICKOLAS BUTLER is the internationally bestselling and prize-winning author of *Shotgun Lovesongs*, *Beneath the Bonfire*, *The Hearts of Men*, *Little Faith*, and *Godspeed*. His work has been translated into more than ten languages and won or been shortlisted for some of France's most prestigious literary prizes. His journalism, poetry, and reviews have appeared in many publications, such as *Ploughshares*, *Narrative*, and *The New York Times*

Book Review to name a few; he also writes a regular column for his local newspaper. Butler lives with his wife and their two children just south of his hometown of Eau Claire, Wisconsin, on sixteen acres of land adjacent to a buffalo farm.

OMAR EL AKKAD was born in Egypt, grew up in Qatar, moved to Canada as a teenager, and now lives in the United States. His fiction and nonfiction writing has appeared in *The Guardian*, *Le Monde*, *Guernica*, *GQ*, and many other newspapers and magazines. His debut novel, *American War*, is an international best seller and has been translated into thirteen languages. It won the Pacific Northwest Booksellers Award, the Oregon Book Award for Fiction, the Kobo Emerging Writer Prize, and has been nominated for more than ten other awards. His second novel, *What Strange Paradise*, was published in 2021 by Knopf.

DELIA FALCONER is the author of two novels, *The Service of Clouds* and *The Lost Thoughts of Soldiers*. Her 2010 nonfiction work, *Sydney*, a personal history of her hometown, won the CAL Waverley Library Award for Literature and was shortlisted for other major national prizes including the New South Wales Premier's History and National Biography awards. She is a senior lecturer in creative writing at University of Technology, Sydney. In 2018 her essay for the *Sydney Review of Books*, "The Opposite of Glamour," won the Walkley-Pascall Award for Arts Criticism.

MELISSA FEBOS is the author of the memoir *Whip Smart* and two essay collections: *Abandon Me*, a Lambda Literary Award finalist and Publishing Triangle Award finalist, and *Girlhood*, a national best seller, which was released on March 30, 2021. Catapult will publish a collection of her craft essays, *Body Work*, in March 2022. The inaugural winner of the Jeanne Córdova Nonfiction Award from Lambda Literary and the recipient of fellowships from the MacDowell Colony, Bread Loaf, Lower Manhattan Cultural Council, the BAU Institute, Vermont Studio Center, the Barbara Deming Foundation, and others, her essays have appeared in *The Paris Review*, *The Believer*, *McSweeney's Quarterly*, *Granta*, *Tin House*, *The Sun*, and *The New York Times*. She is an associate professor at the University of Iowa.

MARY ANNAÏSE HEGLAR is the co-host and co-creator of the Hot Take newsletter and podcast and a climate justice essayist. Her essays about race and climate change have appeared in *Vox*, *Dame*, and other places. She holds a BA in English from Oberlin College and a certificate in editing from New York University.

LACY M. JOHNSON is a Houston-based professor, curator, activist, and author of the essay collection *The Reckonings*, the memoir *The Other Side*—both National Book Critics Circle Award finalists—and the memoir *Trespasses*. Her writing has appeared in *The Best American Essays*, *The Best American Travel Writing*, *The New Yorker*, *The New York Times*, *Los Angeles Times*, *The Paris Review*, *Virginia Quarterly Review*, *Tin House*, *Guernica*, and elsewhere. She teaches creative nonfiction at Rice University and is the founding director of the Houston Flood Museum. In 2020 she was awarded a Guggenheim Fellowship for General Nonfiction.

POROCHISTA KHAKPOUR is the author of the memoir *SICK*, the essay collection *The Brown Album*, and the novels *Tehrangeles*, *The Last Illusion*, and *Sons and Other Flammable Objects*. Among her many fellowships is a National Endowment for the Arts award. Her nonfiction has appeared in *The New York Times*, *Los Angeles Times*, *Elle*, *Bookforum*, *VQR*, and many other publications.

ALEXANDRA KLEEMAN is the author of the novels *Something New Under the Sun* and *You Too Can Have a Body Like Mine*, as well as *Intimations*, a story collection. The winner of a Rome Prize, Berlin Prize, and the Bard Fiction Prize, her work has been published in *The New Yorker*, *The Paris Review*, *Harper's Magazine*, *The New York Times Magazine*, and *n+1*, among other publications. She is an assistant professor at the New School and lives on Staten Island.

LYDIA MILLET has written more than a dozen novels and story collections, often about the ties between people and other animals and the crisis of extinction. Her story collection *Fight No More* received an Award of Merit from the American Academy of Arts and Letters in 2019, and her collection *Love in Infant Monkeys* was a finalist for the Pulitzer Prize in 2010. She also writes essays, opinion pieces, and other ephemera and has worked as an editor and staff writer at the Center for Biological Diversity since 1999. Her latest novel, *A Children's Bible*, was a finalist for the National Book Award for Fiction and one of *The New York Times Book Review*'s Top 10 Books of 2020.

TRACY O'NEILL is the author of *The Hopeful*, one of *Electric Literature*'s Best Novels of 2015, and *Quotients*, a *New York*

Times New & Noteworthy Book, *TOR* Editor's Choice, and *Literary Hub* Favorite Book of 2020. In 2015, she was named a National Book Foundation 5 Under 35 honoree, long-listed for the Flaherty-Dunnan Prize, and was a Narrative Under 30 finalist. In 2012, she was awarded the Center for Fiction's Emerging Writers Fellowship. Her short fiction was distinguished in the Best American Short Stories 2016 and earned a Pushcart Prize nomination in 2017. Her writing has appeared in *Granta*, *Rolling Stone*, *The Atlantic*, *The New Yorker*, *Literary Hub*, *BOMB*, *Vol. 1 Brooklyn*, *The Believer*, *The Literarian*, *The Austin Chronicle*, *New World Writing*, *Narrative*, *Scoundrel Time*, *Guernica*, *Bookforum*, *Electric Literature*, *Grantland*, *Vice*, *The Guardian*, *VQR*, the *San Francisco Chronicle*, and *Catapult*. She holds an MFA from the City College of New York and an MA, MPhil, and PhD from Columbia University. While editor-in-chief of the literary journal *Epiphany*, she established the Breakout 8 Writers Prize with the Authors Guild. She currently teaches at Vassar College.

EMILY RABOTEAU's books are *The Professor's Daughter* and *Searching for Zion*, winner of an American Book Award. Her next book, *Caution: Lessons in Survival*, is forthcoming from Holt. A contributing editor at *Orion Magazine*, and a regular contributor to the *New York Review of Books*, Raboteau teaches creative writing at the City College of New York in Harlem.

RACHEL RIEDERER is a writer from Kansas City, focusing on science, culture, and climate. She is a member of the editorial staff of *The New Yorker*.

KIM STANLEY ROBINSON is an American science fiction writer. He is the author of more than twenty books, including the international best-selling Mars trilogy: *Red Mars*, *Green Mars*, and *Blue Mars*, and more recently *Red Moon*, *New York 2140*, and *2312*, which was a *New York Times* best seller nominated for all seven of the major science fiction awards—a first for any book. In 2008 he was named a "Hero of the Environment" by *Time*, and he works with the Sierra Nevada Research Institute, the Clarion Writers' Workshop, and UC San Diego's Arthur C. Clarke Center for Human Imagination. His latest novel is called *Ministry for the Future.*

ELIZABETH RUSH is the author of *Rising: Dispatches from the New American Shore*, a 2019 finalist for the Pulitzer Prize in General Nonfiction, and *Still Lifes from a Vanishing City: Essays and Photographs from Yangon, Myanmar.* Her writing has been supported by grants from the Alfred P. Sloan Foundation, the National Science Foundation, the Andrew Mellon Foundation, the Howard Foundation, and other organizations. She is currently at work on a book about motherhood and Antarctica's diminishing glaciers. She lives in Providence, where she teaches at Brown University.

MEERA SUBRAMANIAN is an award-winning journalist who has explored the disappearance of India's vultures, questioned the "Good Anthropocene," and investigated perceptions of climate change among conservative Americans. She is a contributing editor of *Orion*, and former MIT Knight Science Journalism fellow and Currie C. and Thomas A. Barron Visiting Professor in the

Environment and the Humanities at Princeton University. Her narrative nonfiction book, *A River Runs Again: India's Natural World in Crisis,* was a 2016 Orion Book Award finalist. You can find her at www.meerasub.org.

PITCHAYA SUDBANTHAD is the author of *Bangkok Wakes to Rain.* The novel, published by Riverhead Books (U.S.) and Sceptre (U.K.), was selected as a notable book of the year by *The New York Times* and *The Washington Post,* as well as a finalist for the Center for Fiction's First Novel Prize. He has received fellowships in fiction writing from the New York Foundation for the Arts and MacDowell, and currently splits time between Bangkok and Brooklyn.

TERESE SVOBODA is a Guggenheim fellow and the author of nineteen books, most recently *Theatrix: Poetry Plays.* She has won the Bobst Prize in fiction, the Iowa Prize for poetry, an NEH grant for translation, the Graywolf Nonfiction Prize, a Jerome Foundation prize for video, the O. Henry award for the short story, a Bobst prize for the novel, and a Pushcart Prize for the essay. She is a three-time winner of the New York Foundation for the Arts fellowship, and has been awarded Headlands, James Merrill, Hawthornden, Yaddo, MacDowell, and Bellagio residencies. Her opera, *WET,* premiered at L.A.'s REDCAT Theater in Disney Hall.

LIDIA YUKNAVITCH is the national best-selling author of the novels *The Book of Joan* and *The Small Backs of Children,* winner of the 2016 Oregon Book Award's Ken Kesey Award for Fiction as well as the Reader's Choice Award, the novel *Dora: A Headcase*, and a critical book on war and narrative, *Allegories of Violence*. Her widely acclaimed memoir *The Chronology of Water* was a finalist for a PEN Center USA award for creative nonfiction and winner of a PNBA Award and the Oregon Book Award Reader's Choice. *The Misfit's Manifesto*, a book based on her recent TED Talk, was published by TED Books. Her new collection of fiction, *Verge*, was published by Riverhead Books in February 2020. She founded the workshop series Corporeal Writing in Portland, Oregon, where she teaches both in person and online. She received her doctorate in literature from the University of Oregon. She lives in Oregon with her husband, Andy Mingo, and their renaissance man son, Miles. She is a very good swimmer.

ABOUT THE EDITORS

AMY BRADY is the executive director of *Orion*. She is also the author of a cultural history of ice in America and the former editor in chief of the *Chicago Review of Books*. She holds a PhD in literature from the University of Massachusetts Amherst and has won writing and research awards from the National Science Foundation, the Bread Loaf Environmental Writers Conference, and the Library of Congress.

TAJJA ISEN is the author of *Some of My Best Friends: Essays on Lip Service*. She is the editor in chief of *Catapult* and the former digital editor at *The Walrus*. Also a voice actor, Isen can be heard on such animated shows as *The Berenstain Bears*, *Atomic Betty*, and *Go Dog Go*, among others.